Use R!

Advisors:
Robert Gentleman
Kurt Hornik
Giovanni Parmigiani

Paul S.P. Cowpertwait · Andrew V. Metcalfe

Introductory Time Series
with R

Paul S.P. Cowpertwait
Inst. Information and
    Mathematical Sciences
Massey University
Auckland
Albany Campus
New Zealand
p.s.cowpertwait@massey.ac.nz

Andrew V. Metcalfe
School of Mathematical
    Sciences
University of Adelaide
Adelaide SA 5005
Australia
andrew.metcalfe@adelaide.edu.au

Series Editors
Robert Gentleman
Program in Computational Biology
Division of Public Health Sciences
Fred Hutchinson Cancer Research Center
1100 Fairview Avenue, N. M2-B876
Seattle, Washington 98109
USA

Kurt Hornik
Department of Statistik and Mathematik
Wirtschaftsuniversität Wien Augasse 2-6
A-1090 Wien
Austria

Giovanni Parmigiani
The Sidney Kimmel Comprehensive Cancer
    Center at Johns Hopkins University
550 North Broadway
Baltimore, MD 21205-2011
USA

ISBN 978-0-387-88697-8   e-ISBN 978-0-387-88698-5
DOI 10.1007/978-0-387-88698-5
Springer Dordrecht Heidelberg London New York

Library of Congress Control Number: 2009928496

Printed on acid-free paper

Springer is part of Springer Science+Business Media (www.springer.com)

In memory of Ian Cowpertwait

Preface

R has a command line interface that offers considerable advantages over menu systems in terms of efficiency and speed once the commands are known and the language understood. However, the command line system can be daunting for the first-time user, so there is a need for concise texts to enable the student or analyst to make progress with R in their area of study. This book aims to fulfil that need in the area of *time series* to enable the non-specialist to progress, at a fairly quick pace, to a level where they can confidently apply a range of time series methods to a variety of data sets. The book assumes the reader has a knowledge typical of a first-year university statistics course and is based around lecture notes from a range of time series courses that we have taught over the last twenty years. Some of this material has been delivered to post-graduate finance students during a concentrated six-week course and was well received, so a selection of the material could be mastered in a concentrated course, although in general it would be more suited to being spread over a complete semester.

The book is based around practical applications and generally follows a similar format for each time series model being studied. First, there is an introductory motivational section that describes practical reasons why the model may be needed. Second, the model is described and defined in mathematical notation. The model is then used to simulate synthetic data using R code that closely reflects the model definition and then fitted to the synthetic data to recover the underlying model parameters. Finally, the model is fitted to an example historical data set and appropriate diagnostic plots given. By using R, the whole procedure can be reproduced by the reader, and it is recommended that students work through most of the examples.[1] Mathematical derivations are provided in separate frames and starred sec-

---

[1] We used the R package Sweave to ensure that, in general, your code will produce the same output as ours. However, for stylistic reasons we sometimes edited our code; e.g., for the plots there will sometimes be minor differences between those generated by the code in the text and those shown in the actual figures.

tions and can be omitted by those wanting to progress quickly to practical applications. At the end of each chapter, a concise summary of the R commands that were used is given followed by exercises. All data sets used in the book, and solutions to the odd numbered exercises, are available on the website http://www.massey.ac.nz/~pscowper/ts.

We thank John Kimmel of Springer and the anonymous referees for their helpful guidance and suggestions, Brian Webby for careful reading of the text and valuable comments, and John Xie for useful comments on an earlier draft. The Institute of Information and Mathematical Sciences at Massey University and the School of Mathematical Sciences, University of Adelaide, are acknowledged for support and funding that made our collaboration possible. Paul thanks his wife, Sarah, for her continual encouragement and support during the writing of this book, and our son, Daniel, and daughters, Lydia and Louise, for the joy they bring to our lives. Andrew thanks Natalie for providing inspiration and her enthusiasm for the project.

Paul Cowpertwait and Andrew Metcalfe

Massey University, Auckland, New Zealand
University of Adelaide, Australia

December 2008

Contents

1

Time Series Data

1.1 Purpose

Time series are analysed to understand the past and to predict the future, enabling managers or policy makers to make properly informed decisions. A time series analysis quantifies the main features in data and the random variation. These reasons, combined with improved computing power, have made time series methods widely applicable in government, industry, and commerce.

The Kyoto Protocol is an amendment to the United Nations Framework Convention on Climate Change. It opened for signature in December 1997 and came into force on February 16, 2005. The arguments for reducing greenhouse gas emissions rely on a combination of science, economics, and time series analysis. Decisions made in the next few years will affect the future of the planet.

During 2006, Singapore Airlines placed an initial order for twenty Boeing 787-9s and signed an order of intent to buy twenty-nine new Airbus planes, twenty A350s, and nine A380s (superjumbos). The airline's decision to expand its fleet relied on a combination of time series analysis of airline passenger trends and corporate plans for maintaining or increasing its market share.

Time series methods are used in everyday operational decisions. For example, gas suppliers in the United Kingdom have to place orders for gas from the offshore fields one day ahead of the supply. Variation about the average for the time of year depends on temperature and, to some extent, the wind speed. Time series analysis is used to forecast demand from the seasonal average with adjustments based on one-day-ahead weather forecasts.

Time series models often form the basis of computer simulations. Some examples are assessing different strategies for control of inventory using a simulated time series of demand; comparing designs of wave power devices using a simulated series of sea states; and simulating daily rainfall to investigate the long-term environmental effects of proposed water management policies.

P.S.P. Cowpertwait and A.V. Metcalfe, *Introductory Time Series with R*,
Use R, DOI 10.1007/978-0-387-88698-5_1,
© Springer Science+Business Media, LLC 2009

1.2 Time series

In most branches of science, engineering, and commerce, there are variables measured sequentially in time. Reserve banks record interest rates and exchange rates each day. The government statistics department will compute the country's gross domestic product on a yearly basis. Newspapers publish yesterday's noon temperatures for capital cities from around the world. Meteorological offices record rainfall at many different sites with differing resolutions. When a variable is measured sequentially in time over or at a fixed interval, known as the *sampling interval*, the resulting data form a *time series*.

Observations that have been collected over fixed sampling intervals form a *historical* time series. In this book, we take a *statistical* approach in which the historical series are treated as realisations of sequences of *random variables*. A sequence of random variables defined at fixed sampling intervals is sometimes referred to as a *discrete-time stochastic process*, though the shorter name *time series model* is often preferred. The theory of stochastic processes is vast and may be studied without necessarily fitting any models to data. However, our focus will be more applied and directed towards model fitting and data analysis, for which we will be using R.[1]

The main features of many time series are trends and seasonal variations that can be modelled deterministically with mathematical functions of time. But, another important feature of most time series is that observations close together in time tend to be correlated (*serially dependent*). Much of the methodology in a time series analysis is aimed at explaining this correlation and the main features in the data using appropriate statistical models and descriptive methods. Once a good model is found and fitted to data, the analyst can use the model to forecast future values, or generate simulations, to guide planning decisions. Fitted models are also used as a basis for statistical tests. For example, we can determine whether fluctuations in monthly sales figures provide evidence of some underlying change in sales that we must now allow for. Finally, a fitted statistical model provides a concise summary of the main characteristics of a time series, which can often be essential for decision makers such as managers or politicians.

Sampling intervals differ in their relation to the data. The data may have been aggregated (for example, the number of foreign tourists arriving per day) or sampled (as in a daily time series of close of business share prices). If data are sampled, the sampling interval must be short enough for the time series to provide a very close approximation to the original continuous signal when it is interpolated. In a volatile share market, close of business prices may not suffice for interactive trading but will usually be adequate to show a company's financial performance over several years. At a quite different timescale,

---

[1] R was initiated by Ihaka and Gentleman (1996) and is an open source implementation of S, a language for data analysis developed at Bell Laboratories (Becker et al. 1988).

time series analysis is the basis for signal processing in telecommunications, engineering, and science. Continuous electrical signals are sampled to provide time series using analog-to-digital (A/D) converters at rates that can be faster than millions of observations per second.

1.3 R language

It is assumed that you have R (version 2 or higher) installed on your computer, and it is suggested that you work through the examples, making sure your output agrees with ours.[2] If you do not have R, then it can be installed free of charge from the Internet site www.r-project.org. It is also recommended that you have some familiarity with the basics of R, which can be obtained by working through the first few chapters of an elementary textbook on R (e.g., Dalgaard 2002) or using the online "An Introduction to R", which is also available via the R help system – type `help.start()` at the command prompt to access this.

R has many features in common with both *functional* and *object oriented* programming languages. In particular, functions in R are treated as objects that can be manipulated or used recursively.[3] For example, the factorial function can be written recursively as

```
> Fact <- function(n) if (n == 1) 1 else n * Fact(n - 1)
> Fact(5)

[1] 120
```

In common with functional languages, assignments in R can be avoided, but they are useful for clarity and convenience and hence will be used in the examples that follow. In addition, R runs faster when 'loops' are avoided, which can often be achieved using matrix calculations instead. However, this can sometimes result in rather obscure-looking code. Thus, for the sake of transparency, loops will be used in many of our examples. Note that R is *case sensitive*, so that X and x, for example, correspond to different variables. In general, we shall use uppercase for the first letter when defining new variables, as this reduces the chance of overwriting inbuilt R functions, which are usually in lowercase.[4]

---

[2] Some of the output given in this book may differ slightly from yours. This is most likely due to editorial changes made for stylistic reasons. For conciseness, we also used `options(digits=3)` to set the number of digits to 4 in the computer output that appears in the book.

[3] Do not be concerned if you are unfamiliar with some of these computing terms, as they are not really essential in understanding the material in this book. The main reason for mentioning them now is to emphasise that R can almost certainly meet your *future* statistical and programming needs should you wish to take the study of time series further.

[4] For example, matrix transpose is `t()`, so t should not be used for *time*.

The best way to learn to do a time series analysis in R is through practice, so we now turn to some examples, which we invite you to work through.

1.4 Plots, trends, and seasonal variation

1.4.1 A flying start: Air passenger bookings

The number of international passenger bookings (in thousands) per month on an airline (Pan Am) in the United States were obtained from the Federal Aviation Administration for the period 1949–1960 (Brown, 1963). The company used the data to predict future demand before ordering new aircraft and training aircrew. The data are available as a time series in R and illustrate several important concepts that arise in an exploratory time series analysis.

Type the following commands in R, and check your results against the output shown here. To save on typing, the data are assigned to a variable called AP.

```
> data(AirPassengers)
> AP <- AirPassengers
> AP
```

```
     Jan Feb Mar Apr May Jun Jul Aug Sep Oct Nov Dec
1949 112 118 132 129 121 135 148 148 136 119 104 118
1950 115 126 141 135 125 149 170 170 158 133 114 140
1951 145 150 178 163 172 178 199 199 184 162 146 166
1952 171 180 193 181 183 218 230 242 209 191 172 194
1953 196 196 236 235 229 243 264 272 237 211 180 201
1954 204 188 235 227 234 264 302 293 259 229 203 229
1955 242 233 267 269 270 315 364 347 312 274 237 278
1956 284 277 317 313 318 374 413 405 355 306 271 306
1957 315 301 356 348 355 422 465 467 404 347 305 336
1958 340 318 362 348 363 435 491 505 404 359 310 337
1959 360 342 406 396 420 472 548 559 463 407 362 405
1960 417 391 419 461 472 535 622 606 508 461 390 432
```

All data in R are stored in *objects*, which have a range of *methods* available. The *class* of an object can be found using the class function:

```
> class(AP)
```

```
[1] "ts"
```

```
> start(AP); end(AP); frequency(AP)
```

```
[1] 1949    1
[1] 1960   12
[1] 12
```

In this case, the object is of class `ts`, which is an abbreviation for 'time series'. Time series objects have a number of *methods* available, which include the functions `start`, `end`, and `frequency` given above. These methods can be listed using the function `methods`, but the output from this function is not always helpful. The key thing to bear in mind is that *generic* functions in R, such as `plot` or `summary`, will attempt to give the most appropriate output to any given input object; try typing `summary(AP)` now to see what happens.

As the objective in this book is to analyse time series, it makes sense to put our data into objects of class `ts`. This can be achieved using a function also called `ts`, but this was not necessary for the airline data, which were already stored in this form. In the next example, we shall create a `ts` object from data read directly from the Internet.

One of the most important steps in a preliminary time series analysis is to plot the data; i.e., create a *time plot*. For a time series object, this is achieved with the generic plot function:

```
> plot(AP, ylab = "Passengers (1000's)")
```

You should obtain a plot similar to Figure 1.1 below. Parameters, such as `xlab` or `ylab`, can be used in `plot` to improve the default labels.

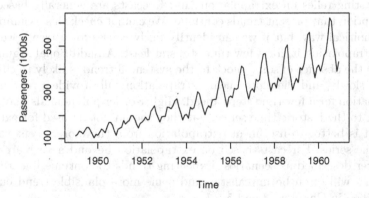

Fig. 1.1. International air passenger bookings in the United States for the period 1949–1960.

There are a number of features in the time plot of the air passenger data that are common to many time series (Fig. 1.1). For example, it is apparent that the number of passengers travelling on the airline is increasing with time. In general, a systematic change in a time series that does not appear to be periodic is known as a *trend*. The simplest model for a trend is a linear increase or decrease, and this is often an adequate approximation.

A repeating pattern within each year is known as *seasonal variation*, although the term is applied more generally to repeating patterns within any fixed period, such as restaurant bookings on different days of the week. There is clear seasonal variation in the air passenger time series. At the time, bookings were highest during the summer months of June, July, and August and lowest during the autumn month of November and winter month of February. Sometimes we may claim there are *cycles* in a time series that do not correspond to some fixed natural period; examples may include business cycles or climatic oscillations such as El Niño. None of these is apparent in the airline bookings time series.

An understanding of the likely causes of the features in the plot helps us formulate an appropriate time series model. In this case, possible causes of the increasing trend include rising prosperity in the aftermath of the Second World War, greater availability of aircraft, cheaper flights due to competition between airlines, and an increasing population. The seasonal variation coincides with vacation periods. In Chapter 5, time series regression models will be specified to allow for underlying causes like these. However, many time series exhibit trends, which might, for example, be part of a longer cycle or be random and subject to unpredictable change. Random, or *stochastic*, trends are common in economic and financial time series. A regression model would not be appropriate for a stochastic trend.

Forecasting relies on extrapolation, and forecasts are generally based on an assumption that present trends continue. We cannot check this assumption in any empirical way, but if we can identify likely causes for a trend, we can justify extrapolating it, for a few time steps at least. An additional argument is that, in the absence of some shock to the system, a trend is likely to change relatively slowly, and therefore linear extrapolation will provide a reasonable approximation for a few time steps ahead. Higher-order polynomials may give a good fit to the historic time series, but they should not be used for extrapolation. It is better to use linear extrapolation from the more recent values in the time series. Forecasts based on extrapolation beyond a year are perhaps better described as scenarios. Expecting trends to continue linearly for many years will often be unrealistic, and some more plausible trend curves are described in Chapters 3 and 5.

A time series plot not only emphasises patterns and features of the data but can also expose *outliers* and *erroneous* values. One cause of the latter is that missing data are sometimes coded using a negative value. Such values need to be handled differently in the analysis and must not be included as observations when fitting a model to data.[5] Outlying values that cannot be attributed to some coding should be checked carefully. If they are correct,

---

[5] Generally speaking, missing values are suitably handled by R, provided they are correctly coded as 'NA'. However, if your data do contain missing values, then it is always worth checking the 'help' on the R function that you are using, as an extra parameter or piece of coding may be required.

they are likely to be of particular interest and should not be excluded from the analysis. However, it may be appropriate to consider *robust methods* of fitting models, which reduce the influence of outliers.

To get a clearer view of the trend, the seasonal effect can be removed by aggregating the data to the annual level, which can be achieved in R using the `aggregate` function. A summary of the values for each season can be viewed using a boxplot, with the `cycle` function being used to extract the seasons for each item of data.

The plots can be put in a single graphics window using the `layout` function, which takes as input a vector (or matrix) for the location of each plot in the display window. The resulting boxplot and annual series are shown in Figure 1.2.

```
> layout(1:2)
> plot(aggregate(AP))
> boxplot(AP ~ cycle(AP))
```

You can see an increasing trend in the annual series (Fig. 1.2a) and the seasonal effects in the boxplot. More people travelled during the summer months of June to September (Fig. 1.2b).

1.4.2 Unemployment: Maine

Unemployment rates are one of the main economic indicators used by politicians and other decision makers. For example, they influence policies for regional development and welfare provision. The monthly unemployment rate for the US state of Maine from January 1996 until August 2006 is plotted in the upper frame of Figure 1.3. In any time series analysis, it is essential to understand how the data have been collected and their unit of measurement. The US Department of Labor gives precise definitions of terms used to calculate the unemployment rate.

The monthly unemployment data are available in a file online that is read into R in the code below. Note that the first row in the file contains the name of the variable (`unemploy`), which can be accessed directly once the `attach` command is given. Also, the `header` parameter must be set to `TRUE` so that R treats the first row as the variable name rather than data.

```
> www <- "http://www.massey.ac.nz/~pscowper/ts/Maine.dat"
> Maine.month <- read.table(www, header = TRUE)

> attach(Maine.month)
> class(Maine.month)

[1] "data.frame"
```

When we read data in this way from an ASCII text file, the 'class' is not time series but `data.frame`. The `ts` function is used to convert the data to a time series object. The following command creates a time series object:

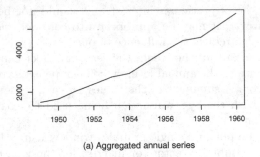

(a) Aggregated annual series

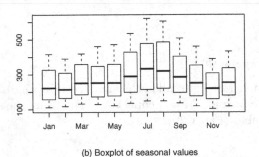

(b) Boxplot of seasonal values

Fig. 1.2. International air passenger bookings in the United States for the period 1949–1960. Units on the y-axis are 1000s of people. (a) Series aggregated to the annual level; (b) seasonal boxplots of the data.

```
> Maine.month.ts <- ts(unemploy, start = c(1996, 1), freq = 12)
```

This uses all the data. You can select a smaller number by specifying an earlier end date using the parameter `end`. If we wish to analyse trends in the unemployment rate, annual data will suffice. The average (mean) over the twelve months of each year is another example of aggregated data, but this time we divide by 12 to give a mean annual rate.

```
> Maine.annual.ts <- aggregate(Maine.month.ts)/12
```

We now plot both time series. There is clear monthly variation. From Figure 1.3(a) it seems that the February figure is typically about 20% more than the annual average, whereas the August figure tends to be roughly 20% less.

```
> layout(1:2)
> plot(Maine.month.ts, ylab = "unemployed (%)")
> plot(Maine.annual.ts, ylab = "unemployed (%)")
```

We can calculate the precise percentages in R, using `window`. This function will extract that part of the time series between specified start and end points

and will sample with an interval equal to `frequency` if its argument is set to
TRUE. So, the first line below gives a time series of February figures.

```
> Maine.Feb <- window(Maine.month.ts, start = c(1996,2), freq = TRUE)
> Maine.Aug <- window(Maine.month.ts, start = c(1996,8), freq = TRUE)
> Feb.ratio <- mean(Maine.Feb) / mean(Maine.month.ts)
> Aug.ratio <- mean(Maine.Aug) / mean(Maine.month.ts)

> Feb.ratio
[1] 1.223
> Aug.ratio
[1] 0.8164
```

On average, unemployment is 22% higher in February and 18% lower in
August. An explanation is that Maine attracts tourists during the summer,
and this creates more jobs. Also, the period before Christmas and over the
New Year's holiday tends to have higher employment rates than the first few
months of the new year. The annual unemployment rate was as high as 8.5%
in 1976 but was less than 4% in 1988 and again during the three years 1999–
2001. If we had sampled the data in August of each year, for example, rather
than taken yearly averages, we would have consistently underestimated the
unemployment rate by a factor of about 0.8.

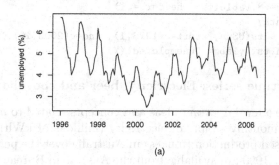

(a)

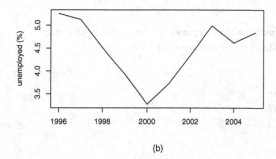

(b)

Fig. 1.3. Unemployment in Maine: (a) monthly January 1996–August 2006; (b)
annual 1996–2005.

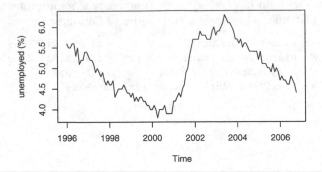

Fig. 1.4. Unemployment in the United States January 1996–October 2006.

The monthly unemployment rate for all of the United States from January 1996 until October 2006 is plotted in Figure 1.4. The decrease in the unemployment rate around the millennium is common to Maine and the United States as a whole, but Maine does not seem to be sharing the current US decrease in unemployment.

```
> www <- "http://www.massey.ac.nz/~pscowper/ts/USunemp.dat"
> US.month <- read.table(www, header = T)
> attach(US.month)
> US.month.ts <- ts(USun, start=c(1996,1), end=c(2006,10), freq = 12)
> plot(US.month.ts, ylab = "unemployed (%)")
```

1.4.3 Multiple time series: Electricity, beer and chocolate data

Here we illustrate a few important ideas and concepts related to *multiple* time series data. The monthly supply of electricity (millions of kWh), beer (Ml), and chocolate-based production (tonnes) in Australia over the period January 1958 to December 1990 are available from the Australian Bureau of Statistics (ABS).[6] The three series have been stored in a single file online, which can be read as follows:

```
www <- "http://www.massey.ac.nz/~pscowper/ts/cbe.dat"
CBE <- read.table(www, header = T)

> CBE[1:4, ]

  choc beer elec
1 1451 96.3 1497
2 2037 84.4 1463
3 2477 91.2 1648
4 2785 81.9 1595
```

---

[6] ABS data used with permission from the Australian Bureau of Statistics: http://www.abs.gov.au.

```
> class(CBE)
```

```
[1] "data.frame"
```

Now create time series objects for the electricity, beer, and chocolate data. If you omit end, R uses the full length of the vector, and if you omit the month in start, R assumes 1. You can use plot with cbind to plot several series on one figure (Fig. 1.5).

```
> Elec.ts <- ts(CBE[, 3], start = 1958, freq = 12)
> Beer.ts <- ts(CBE[, 2], start = 1958, freq = 12)
> Choc.ts <- ts(CBE[, 1], start = 1958, freq = 12)
> plot(cbind(Elec.ts, Beer.ts, Choc.ts))
```

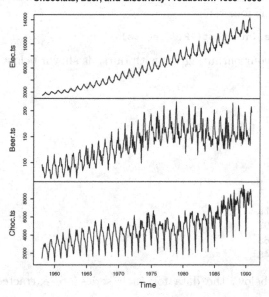

Fig. 1.5. Australian chocolate, beer, and electricity production; January 1958–December 1990.

The plots in Figure 1.5 show increasing trends in production for all three goods, partly due to the rising population in Australia from about 10 million to about 18 million over the same period (Fig. 1.6). But notice that electricity production has risen by a factor of 7, and chocolate production by a factor of 4, over this period during which the population has not quite doubled.

The three series constitute a *multiple* time series. There are many functions in R for handling more than one series, including ts.intersect to obtain the intersection of two series that overlap in time. We now illustrate the use of the intersect function and point out some potential pitfalls in analysing multiple

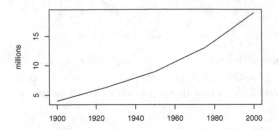

Fig. 1.6. Australia's population, 1900–2000.

time series. The intersection between the air passenger data and the electricity data is obtained as follows:

```
> AP.elec <- ts.intersect(AP, Elec.ts)
```

Now check that your output agrees with ours, as shown below.

```
> start(AP.elec)

[1] 1958    1

> end(AP.elec)

[1] 1960   12

> AP.elec[1:3, ]

       AP Elec.ts
[1,] 340    1497
[2,] 318    1463
[3,] 362    1648
```

In the code below, the data for each series are extracted and plotted (Fig. 1.7).[7]

```
> AP <- AP.elec[,1]; Elec <- AP.elec[,2]

> layout(1:2)
> plot(AP,   main = "", ylab = "Air passengers / 1000's")
> plot(Elec, main = "", ylab = "Electricity production / MkWh")

> plot(as.vector(AP), as.vector(Elec),
                  xlab = "Air passengers / 1000's",
                  ylab = "Electricity production / MWh")
> abline(reg = lm(Elec ~ AP))
```

---

[7] R is case sensitive, so lowercase is used here to represent the shorter record of air passenger data. In the code, we have also used the argument main="" to suppress unwanted titles.

```
> cor(AP, Elec)

[1] 0.884
```

In the `plot` function above, `as.vector` is needed to convert the `ts` objects to
ordinary vectors suitable for a scatter plot.

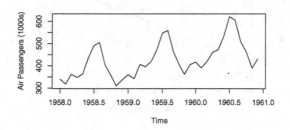

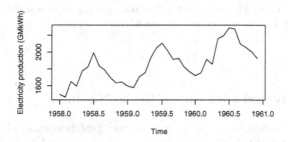

Fig. 1.7. International air passengers and Australian electricity production for the
period 1958–1960. The plots look similar because both series have an increasing
trend and a seasonal cycle. However, this does not imply that there exists a causal
relationship between the variables.

The two time series are highly correlated, as can be seen in the plots, with a
correlation coefficient of 0.88. Correlation will be discussed more in Chapter 2,
but for the moment observe that the two time plots look similar (Fig. 1.7) and
that the scatter plot shows an approximate linear association between the two
variables (Fig. 1.8). However, it is important to realise that correlation does
not imply causation. In this case, it is not plausible that higher numbers of
air passengers in the United States cause, or are caused by, higher electricity
production in Australia. A reasonable explanation for the correlation is that
the increasing prosperity and technological development in both countries over
this period accounts for the increasing trends. The two time series also happen
to have similar seasonal variations. For these reasons, it is usually appropriate
to remove trends and seasonal effects before comparing multiple series. This
is often achieved by working with the residuals of a regression model that has
deterministic terms to represent the trend and seasonal effects (Chapter 5).

In the simplest cases, the residuals can be modelled as independent random variation from a single distribution, but much of the book is concerned with fitting more sophisticated models.

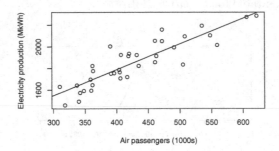

Fig. 1.8. Scatter plot of air passengers and Australian electricity production for the period: 1958–1960. The apparent linear relationship between the two variables is misleading and a consequence of the trends in the series.

1.4.4 Quarterly exchange rate: GBP to NZ dollar

The trends and seasonal patterns in the previous two examples were clear from the plots. In addition, reasonable explanations could be put forward for the possible causes of these features. With financial data, exchange rates for example, such marked patterns are less likely to be seen, and different methods of analysis are usually required. A financial series may sometimes show a dramatic change that has a clear cause, such as a war or natural disaster. Day-to-day changes are more difficult to explain because the underlying causes are complex and impossible to isolate, and it will often be unrealistic to assume any deterministic component in the time series model.

The exchange rates for British pounds sterling to New Zealand dollars for the period January 1991 to March 2000 are shown in Figure 1.9. The data are mean values taken over *quarterly* periods of three months, with the first quarter being January to March and the last quarter being October to December. They can be read into R from the book website and converted to a quarterly time series as follows:

```
> www <- "http://www.massey.ac.nz/~pscowper/ts/pounds_nz.dat"
> Z <- read.table(www, header = T)

> Z[1:4, ]

[1] 2.92 2.94 3.17 3.25

> Z.ts <- ts(Z, st = 1991, fr = 4)
```

```
> plot(Z.ts, xlab = "time / years",
              ylab = "Quarterly exchange rate in $NZ / pound")
```

Short-term trends are apparent in the time series: After an initial surge ending in 1992, a negative trend leads to a minimum around 1996, which is followed by a positive trend in the second half of the series (Fig. 1.9).

The trend seems to change direction at unpredictable times rather than displaying the relatively consistent pattern of the air passenger series and Australian production series. Such trends have been termed *stochastic trends* to emphasise this randomness and to distinguish them from more *deterministic* trends like those seen in the previous examples. A mathematical model known as a *random walk* can sometimes provide a good fit to data like these and is fitted to this series in §4.4.2. Stochastic trends are common in financial series and will be studied in more detail in Chapters 4 and 7.

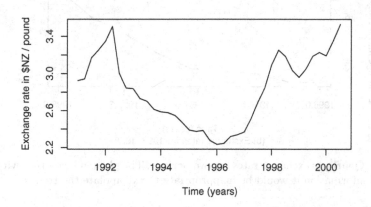

Fig. 1.9. Quarterly exchange rates for the period 1991–2000.

Two local trends are emphasised when the series is partitioned into two subseries based on the periods 1992–1996 and 1996–1998. The `window` function can be used to extract the subseries:

```
> Z.92.96 <- window(Z.ts, start = c(1992, 1), end = c(1996, 1))
> Z.96.98 <- window(Z.ts, start = c(1996, 1), end = c(1998, 1))

> layout (1:2)
> plot(Z.92.96, ylab = "Exchange rate in $NZ/pound",
                xlab = "Time (years)" )
> plot(Z.96.98, ylab = "Exchange rate in $NZ/pound",
                xlab = "Time (years)" )
```

Now suppose we were observing this series at the start of 1992; i.e., we had the data in Figure 1.10(a). It might have been tempting to predict a

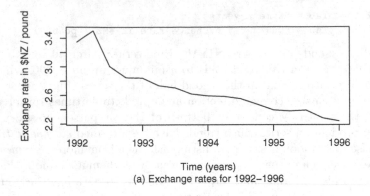

(a) Exchange rates for 1992–1996

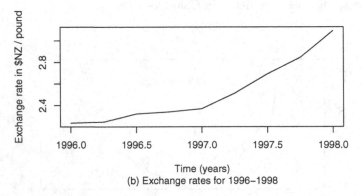

(b) Exchange rates for 1996–1998

Fig. 1.10. Quarterly exchange rates for two periods. The plots indicate that without additional information it would be inappropriate to extrapolate the trends.

continuation of the downward trend for future years. However, this would have been a very poor prediction, as Figure 1.10(b) shows that the data started to follow an increasing trend. Likewise, without additional information, it would also be inadvisable to extrapolate the trend in Figure 1.10(b). This illustrates the potential pitfall of inappropriate extrapolation of stochastic trends when underlying causes are not properly understood. To reduce the risk of making an inappropriate forecast, statistical tests, introduced in Chapter 7, can be used to test for a stochastic trend.

1.4.5 Global temperature series

A change in the world's climate will have a major impact on the lives of many people, as global warming is likely to lead to an increase in ocean levels and natural hazards such as floods and droughts. It is likely that the world economy will be severely affected as governments from around the globe try to enforce a reduction in fossil fuel use and measures are taken to deal with any increase in natural disasters.[8]

In climate change studies (e.g., see Jones and Moberg, 2003; Rayner et al. 2003), the following global temperature series, expressed as anomalies from the monthly means over the period 1961–1990, plays a central role:[9]

```
> www <- "http://www.massey.ac.nz/~pscowper/ts/global.dat"
> Global <- scan(www)
> Global.ts <- ts(Global, st = c(1856, 1), end = c(2005, 12),
    fr = 12)
> Global.annual <- aggregate(Global.ts, FUN = mean)
> plot(Global.ts)
> plot(Global.annual)
```

It is the trend that is of most concern, so the **aggregate** function is used to remove any seasonal effects within each year and produce an annual series of mean temperatures for the period 1856 to 2005 (Fig. 1.11b). We can avoid explicitly dividing by 12 if we specify FUN=mean in the **aggregate** function.

The upward trend from about 1970 onwards has been used as evidence of global warming (Fig. 1.12). In the code below, the monthly time intervals corresponding to the 36-year period 1970–2005 are extracted using the **time** function and the associated observed temperature series extracted using **window**. The data are plotted and a line superimposed using a regression of temperature on the new time index (Fig. 1.12).

```
> New.series <- window(Global.ts, start=c(1970, 1), end=c(2005, 12))
> New.time <- time(New.series)
> plot(New.series); abline(reg=lm(New.series ~ New.time))
```

In the previous section, we discussed a potential pitfall of inappropriate extrapolation. In climate change studies, a vital question is whether rising temperatures are a consequence of human activity, specifically the burning of fossil fuels and increased greenhouse gas emissions, or are a natural trend, perhaps part of a longer cycle, that may decrease in the future without needing a global reduction in the use of fossil fuels. We cannot attribute the increase in global temperature to the increasing use of fossil fuels without invoking some physical explanation[10] because, as we noted in §1.4.3, two unrelated time series will be correlated if they both contain a trend. However, as the general consensus among scientists is that the trend in the global temperature series is related to a global increase in greenhouse gas emissions, it seems reasonable to

---

[8] For general policy documents and discussions on climate change, see the website (and links) for the United Nations Framework Convention on Climate Change at http://unfccc.int.

[9] The data are updated regularly and can be downloaded free of charge from the Internet at: http://www.cru.uea.ac.uk/cru/data/.

[10] For example, refer to US Energy Information Administration at http://www.eia.doe.gov/emeu/aer/inter.html.

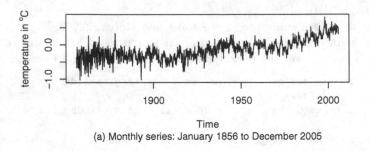

Time
(a) Monthly series: January 1856 to December 2005

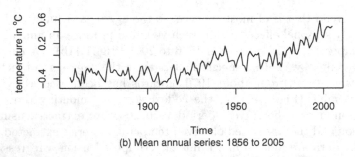

Time
(b) Mean annual series: 1856 to 2005

Fig. 1.11. Time plots of the global temperature series (°C).

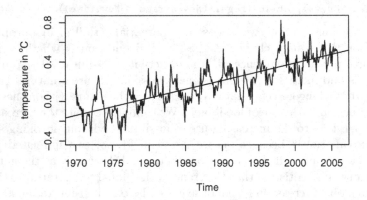

Time

Fig. 1.12. Rising mean global temperatures, January 1970–December 2005. According to the United Nations Framework Convention on Climate Change, the mean global temperature is expected to continue to rise in the future unless greenhouse gas emissions are reduced on a global scale.

acknowledge a causal relationship and to expect the mean global temperature to continue to rise if greenhouse gas emissions are not reduced.[11]

1.5 Decomposition of series

1.5.1 Notation

So far, our analysis has been restricted to plotting the data and looking for features such as trend and seasonal variation. This is an important first step, but to progress we need to fit time series models, for which we require some notation. We represent a time series of length n by $\{x_t : t = 1, \ldots, n\} = \{x_1, x_2, \ldots, x_n\}$. It consists of n values sampled at discrete times $1, 2, \ldots, n$. The notation will be abbreviated to $\{x_t\}$ when the length n of the series does not need to be specified. The time series model is a sequence of random variables, and the observed time series is considered a realisation from the model. We use the same notation for both and rely on the context to make the distinction.[12] An overline is used for sample means:

$$\bar{x} = \sum x_i/n \qquad (1.1)$$

The 'hat' notation will be used to represent a *prediction* or *forecast*. For example, with the series $\{x_t : t = 1, \ldots, n\}$, $\hat{x}_{t+k|t}$ is a *forecast* made at time t for a future value at time $t + k$. A forecast is a predicted future value, and the number of time steps into the future is the *lead time* (k). Following our convention for time series notation, $\hat{x}_{t+k|t}$ can be the random variable or the numerical value, depending on the context.

1.5.2 Models

As the first two examples showed, many series are dominated by a trend and/or seasonal effects, so the models in this section are based on these components. A simple *additive decomposition* model is given by

$$x_t = m_t + s_t + z_t \qquad (1.2)$$

where, at time t, x_t is the observed series, m_t is the trend, s_t is the seasonal effect, and z_t is an error term that is, in general, a sequence of correlated random variables with mean zero. In this section, we briefly outline two main approaches for extracting the trend m_t and the seasonal effect s_t in Equation (1.2) and give the main R functions for doing this.

---

[11] Refer to http://unfccc.int.
[12] Some books do distinguish explicitly by using lowercase for the time series and uppercase for the model.

If the seasonal effect tends to increase as the trend increases, a multiplicative model may be more appropriate:

$$x_t = m_t \cdot s_t + z_t \qquad (1.3)$$

If the random variation is modelled by a multiplicative factor and the variable is positive, an additive decomposition model for $\log(x_t)$ can be used:[13]

$$\log(x_t) = m_t + s_t + z_t \qquad (1.4)$$

Some care is required when the exponential function is applied to the predicted mean of $\log(x_t)$ to obtain a prediction for the mean value x_t, as the effect is usually to bias the predictions. If the random series z_t are normally distributed with mean 0 and variance σ^2, then the predicted mean value at time t based on Equation (1.4) is given by

$$\hat{x}_t = e^{m_t + s_t} e^{\frac{1}{2}\sigma^2} \qquad (1.5)$$

However, if the error series is not normally distributed and is negatively skewed,[14] as is often the case after taking logarithms, the bias correction factor will be an overcorrection (Exercise 4) and it is preferable to apply an empirical adjustment (which is discussed further in Chapter 5). The issue is of practical importance. For example, if we make regular financial forecasts without applying an adjustment, we are likely to consistently underestimate mean costs.

1.5.3 Estimating trends and seasonal effects

There are various ways to estimate the trend m_t at time t, but a relatively simple procedure, which is available in R and does not assume any specific form is to calculate a *moving average* centred on x_t. A moving average is an average of a specified number of time series values around each value in the time series, with the exception of the first few and last few terms. In this context, the length of the moving average is chosen to average out the seasonal effects, which can be estimated later. For monthly series, we need to average twelve consecutive months, but there is a slight snag. Suppose our time series begins at January ($t = 1$) and we average January up to December ($t = 12$). This average corresponds to a time $t = 6.5$, between June and July. When we come to estimate seasonal effects, we need a moving average at integer times. This can be achieved by averaging the average of January up to December and the average of February ($t = 2$) up to January ($t = 13$). This average of

---

[13] To be consistent with R, we use log for the natural logarithm, which is often written ln.

[14] A probability distribution is negatively skewed if its density has a long tail to the left.

two moving averages corresponds to $t = 7$, and the process is called centring. Thus the trend at time t can be estimated by the centred moving average

$$\hat{m}_t = \frac{\frac{1}{2}x_{t-6} + x_{t-5} + \ldots + x_{t-1} + x_t + x_{t+1} + \ldots + x_{t+5} + \frac{1}{2}x_{t+6}}{12} \quad (1.6)$$

where $t = 7, \ldots, n - 6$. The coefficients in Equation (1.6) for each month are $1/12$ (or sum to $1/12$ in the case of the first and last coefficients), so that equal weight is given to each month and the coefficients sum to 1. By using the seasonal frequency for the coefficients in the moving average, the procedure generalises for any seasonal frequency (e.g., quarterly series), provided the condition that the coefficients sum to unity is still met.

An estimate of the monthly additive effect (s_t) at time t can be obtained by subtracting $\hat{m}_t$:

$$\hat{s}_t = x_t - \hat{m}_t \quad (1.7)$$

By averaging these estimates of the monthly effects for each month, we obtain a single estimate of the effect for each month. If the period of the time series is a whole number of years, the number of monthly effects averaged for each month is one less than the number of years of record. At this stage, the twelve monthly additive components should have an average value close to, but not usually exactly equal to, zero. It is usual to adjust them by subtracting this mean so that they do average zero. If the monthly effect is multiplicative, the estimate is given by division; i.e., $\hat{s}_t = x_t/\hat{m}_t$. It is usual to adjust monthly multiplicative factors so that they average unity. The procedure generalises, using the same principle, to any seasonal frequency.

It is common to present economic indicators, such as unemployment percentages, as *seasonally adjusted* series. This highlights any trend that might otherwise be masked by seasonal variation attributable, for instance, to the end of the academic year, when school and university leavers are seeking work. If the seasonal effect is additive, a seasonally adjusted series is given by $x_t - \bar{s}_t$, whilst if it is multiplicative, an adjusted series is obtained from $x_t/\bar{s}_t$, where $\bar{s}_t$ is the seasonally adjusted mean for the month corresponding to time t.

1.5.4 Smoothing

The centred moving average is an example of a *smoothing* procedure that is applied retrospectively to a time series with the objective of identifying an underlying signal or trend. Smoothing procedures can, and usually do, use points before and after the time at which the smoothed estimate is to be calculated. A consequence is that the smoothed series will have some points missing at the beginning and the end unless the smoothing algorithm is adapted for the end points.

A second smoothing algorithm offered by R is stl. This uses a locally weighted regression technique known as *loess*. The regression, which can be a line or higher polynomial, is referred to as local because it uses only some relatively small number of points on either side of the point at which the smoothed estimate is required. The weighting reduces the influence of outlying points and is an example of robust regression. Although the principles behind stl are straightforward, the details are quite complicated.

Smoothing procedures such as the centred moving average and loess do not require a predetermined model, but they do not produce a formula that can be extrapolated to give forecasts. Fitting a line to model a linear trend has an advantage in this respect.

The term *filtering* is also used for smoothing, particularly in the engineering literature. A more specific use of the term filtering is the process of obtaining the best estimate of some variable now, given the latest measurement of it and past measurements. The measurements are subject to random error and are described as being *corrupted by noise*. Filtering is an important part of control algorithms which have a myriad of applications. An exotic example is the Huygens probe leaving the Cassini orbiter to land on Saturn's largest moon, Titan, on January 14, 2005.

1.5.5 Decomposition in R

In R, the function decompose estimates trends and seasonal effects using a moving average method. Nesting the function within plot (e.g., using plot(stl())) produces a single figure showing the original series x_t and the decomposed series m_t, s_t, and z_t. For example, with the electricity data, additive and multiplicative decomposition plots are given by the commands below; the last plot, which uses lty to give different line types, is the superposition of the seasonal effect on the trend (Fig. 1.13).

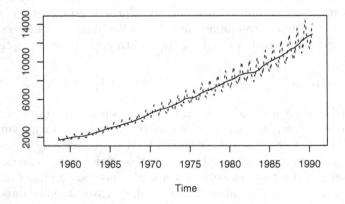

Fig. 1.13. Electricity production data: trend with superimposed multiplicative seasonal effects.

```
> plot(decompose(Elec.ts))
> Elec.decom <- decompose(Elec.ts, type = "mult")
> plot(Elec.decom)
> Trend <- Elec.decom$trend
> Seasonal <- Elec.decom$seasonal
> ts.plot(cbind(Trend, Trend * Seasonal), lty = 1:2)
```

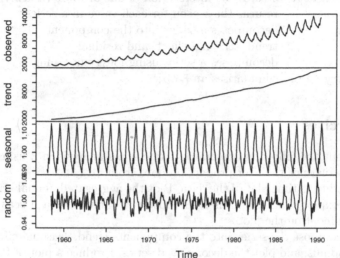

Fig. 1.14. Decomposition of the electricity production data.

In this example, the multiplicative model would seem more appropriate than the additive model because the variance of the original series and trend increase with time (Fig. 1.14). However, the random component, which corresponds to z_t, also has an increasing variance, which indicates that a log-transformation (Equation (1.4)) may be more appropriate for this series (Fig. 1.14). The `random` series obtained from the `decompose` function is not precisely a realisation of the random process z_t but rather an estimate of that realisation. It is an estimate because it is obtained from the original time series using estimates of the trend and seasonal effects. This estimate of the realisation of the random process is a *residual error series*. However, we treat it as a realisation of the random process.

There are many other reasonable methods for decomposing time series, and we cover some of these in Chapter 5 when we study regression methods.

1.6 Summary of commands used in examples

read.table	reads data into a data frame
attach	makes names of column variables available
ts	produces a time series object
aggregate	creates an aggregated series
ts.plot	produces a time plot for one or more series
window	extracts a subset of a time series
time	extracts the time from a time series object
ts.intersect	creates the intersection of one or more time series
cycle	returns the season for each value in a series
decompose	decomposes a series into the components trend, seasonal effect, and residual
stl	decomposes a series using loess smoothing
summary	summarises an R object

1.7 Exercises

1. Carry out the following exploratory time series analysis in R using either the chocolate or the beer production data from §1.4.3.
   a) Produce a time plot of the data. Plot the aggregated annual series and a boxplot that summarises the observed values for each season, and comment on the plots.
   b) Decompose the series into the components trend, seasonal effect, and residuals, and plot the decomposed series. Produce a plot of the trend with a superimposed seasonal effect.

2. Many economic time series are based on indices. A price index is the ratio of the cost of a basket of goods now to its cost in some base year. In the Laspeyre formulation, the basket is based on typical purchases in the base year. You are asked to calculate an index of motoring cost from the following data. The clutch represents all mechanical parts, and the quantity allows for this.

item (i)	quantity '00 (q_{i0})	unit price '00 (p_{i0})	quantity '04 (q_{it})	unit price '04 (p_{it})
car	0.33	18 000	0.5	20 000
petrol (litre)	2 000	0.80	1 500	1.60
servicing (h)	40	40	20	60
tyre	3	80	2	120
clutch	2	200	1	360

The *Laspeyre Price Index* at time t relative to base year 0 is

$$LI_t = \frac{\sum q_{i0}p_{it}}{\sum q_{i0}p_{i0}}$$

Calculate the LI_t for 2004 relative to 2000.

3. The *Paasche Price Index* at time t relative to base year 0 is

$$PI_t = \frac{\sum q_{it} p_{it}}{\sum q_{it} p_{i0}}$$

   a) Use the data above to calculate the PI_t for 2004 relative to 2000.
   b) Explain why the PI_t is usually lower than the LI_t.
   c) Calculate the *Irving-Fisher Price Index* as the geometric mean of LI_t and PI_t. (The geometric mean of a sample of n items is the nth root of their product.)

4. A standard procedure for finding an approximate mean and variance of a function of a variable is to use a Taylor expansion for the function about the mean of the variable. Suppose the variable is y and that its mean and standard deviation are μ and σ respectively.

$$\phi(y) = \phi(\mu) + \phi'(\mu)(y - \mu) + \phi''(\mu)\frac{(y-\mu)^2}{2!} + \phi'''(\mu)\frac{(y-\mu)^3}{3!} + \dots$$

Consider the case of $\phi(.)$ as $e^{(.)}$. By taking the expectation of both sides of this equation, explain why the bias correction factor given in Equation (1.5) is an overcorrection if the residual series has a negative skewness, where the *skewness* γ of a random variable y is defined by

$$\gamma = \frac{E\left[(y-\mu)^3\right]}{\sigma^3}$$

2

Correlation

2.1 Purpose

Once we have identified any trend and seasonal effects, we can deseasonalise
the time series and remove the trend. If we use the additive decomposition
method of §1.5, we first calculate the seasonally adjusted time series and
then remove the trend by subtraction. This leaves the random component,
but the random component is not necessarily well modelled by independent
random variables. In many cases, consecutive variables will be correlated. If
we identify such correlations, we can improve our forecasts, quite dramatically
if the correlations are high. We also need to estimate correlations if we are
to generate realistic time series for simulations. The correlation structure of a
time series model is defined by the correlation function, and we estimate this
from the observed time series.

Plots of serial correlation (the 'correlogram', defined later) are also used
extensively in signal processing applications. The paradigm is an underlying
deterministic signal corrupted by noise. Signals from yachts, ships, aeroplanes,
and space exploration vehicles are examples. At the beginning of 2007, NASA's
twin Voyager spacecraft were sending back radio signals from the frontier of
our solar system, including evidence of hollows in the turbulent zone near the
edge.

2.2 Expectation and the ensemble

2.2.1 Expected value

The *expected value*, commonly abbreviated to *expectation*, E, of a variable,
or a function of a variable, is its *mean* value in a population. So $E(x)$ is the
mean of x, denoted μ,[1] and $E\left[(x - \mu)^2\right]$ is the mean of the squared deviations

---

[1] A more formal definition of the expectation E of a function $\phi(x, y)$ of continuous
random variables x and y, with a joint probability density function $f(x, y)$, is the

P.S.P. Cowpertwait and A.V. Metcalfe, *Introductory Time Series with R*,
Use R, DOI 10.1007/978-0-387-88698-5_2,
© Springer Science+Business Media, LLC 2009

about μ, better known as the *variance* σ^2 of x.[2] The standard deviation, σ is the square root of the variance. If there are two variables (x, y), the variance may be generalised to the *covariance*, $\gamma(x, y)$. Covariance is defined by

$$\gamma(x, y) = E\left[(x - \mu_x)(y - \mu_y)\right] \tag{2.1}$$

The covariance is a measure of *linear association* between two variables (x, y). In §1.4.3, we emphasised that a linear association between variables does not imply causality.

Sample estimates are obtained by adding the appropriate function of the individual data values and division by n or, in the case of variance and covariance, $n - 1$, to give unbiased estimators.[3] For example, if we have n data pairs, (x_i, y_i), the sample covariance is given by

$$\text{Cov}(x, y) = \sum (x_i - \overline{x})(y_i - \overline{y})/(n - 1) \tag{2.2}$$

If the data pairs are plotted, the lines $x = \overline{x}$ and $y = \overline{y}$ divide the plot into quadrants. Points in the lower left quadrant have both $(x_i - \overline{x})$ and $(y_i - \overline{y})$ negative, so the product that contributes to the covariance is positive. Points in the upper right quadrant also make a positive contribution. In contrast, points in the upper left and lower right quadrants make a negative contribution to the covariance. Thus, if y tends to increase when x increases, most of the points will be in the lower left and upper right quadrants and the covariance will be positive. Conversely, if y tends to decrease as x increases, the covariance will be negative. If there is no such linear association, the covariance will be small relative to the standard deviations of $\{x_i\}$ and $\{y_i\}$ – always check the plot in case there is a quadratic association or some other pattern. In R we can calculate a sample covariance, with denominator $n - 1$, from its definition or by using the function cov. If we use the **mean** function, we are implicitly dividing by n.

Benzoapyrene is a carcinogenic hydrocarbon that is a product of incomplete combustion. One source of benzoapyrene and carbon monoxide is automobile exhaust. Colucci and Begeman (1971) analysed sixteen air samples

---

mean value for ϕ obtained by integrating over all possible values of x and y:

$$E\left[\phi(x, y)\right] = \int_y \int_x \phi(x, y) f(x, y)\, dx\, dy$$

Note that the mean of x is obtained as the special case $\phi(x, y) = x$.

[2] For more than one variable, subscripts can be used to distinguish between the properties; e.g., for the means we may write μ_x and μ_y to distinguish between the mean of x and the mean of y.

[3] An estimator is unbiased for a population parameter if its average value, in infinitely repeated samples of size n, equals that population parameter. If an estimator is unbiased, its value in a particular sample is referred to as an unbiased estimate.

from Herald Square in Manhattan and recorded the carbon monoxide concentration (x, in parts per million) and benzoapyrene concentration (y, in micrograms per thousand cubic metres) for each sample. The data are plotted in Figure 2.1.

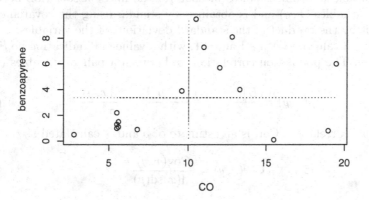

Fig. 2.1. Sixteen air samples from Herald Square.

```
> www <- "http://www.massey.ac.nz/~pscowper/ts/Herald.dat"
> Herald.dat <- read.table(www, header = T)
> attach (Herald.dat)
```

We now use R to calculate the covariance for the Herald Square pairs in three different ways:

```
> x <- CO; y <- Benzoa; n <- length(x)
> sum((x - mean(x))*(y - mean(y))) / (n - 1)

[1] 5.51

> mean((x - mean(x)) * (y - mean(y)))

[1] 5.17

> cov(x, y)

[1] 5.51
```

The correspondence between the R code above and the expectation definition of covariance should be noted:

$$\texttt{mean((x - mean(x))*(y - mean(y)))} \rightarrow E\left[(x - \mu_x)(y - \mu_y)\right] \quad (2.3)$$

Given this correspondence, the more natural estimate of covariance would be mean((x - mean(x))*(y - mean(y))). However, as can be seen above, the values computed using the internal function cov are those obtained using sum with a denominator of $n - 1$. As n gets large, the difference in denominators becomes less noticeable and the more natural estimate *asymptotically* approaches the unbiased estimate.[4]

Correlation is a dimensionless measure of the linear association between a pair of variables (x, y) and is obtained by standardising the covariance by dividing it by the product of the standard deviations of the variables. Correlation takes a value between -1 and $+1$, with a value of 0 indicating no *linear* association. The population correlation, ρ, between a pair of variables (x, y) is defined by

$$\rho(x, y) = \frac{E\left[(x - \mu_x)(y - \mu_y)\right]}{\sigma_x \sigma_y} = \frac{\gamma(x, y)}{\sigma_x \sigma_y} \qquad (2.4)$$

The sample correlation, Cor, is an estimate of ρ and is calculated as

$$\mathrm{Cor}(x, y) = \frac{\mathrm{Cov}(x, y)}{\mathrm{sd}(x)\mathrm{sd}(y)} \qquad (2.5)$$

In R, the sample correlation for pairs (x_i, y_i) stored in vectors x and y is cor(x,y). A value of $+1$ or -1 indicates an exact linear association, with the (x, y) pairs falling on a straight line of positive or negative slope, respectively. The correlation between the CO and benzoapyrene measurements at Herald Square is now calculated both from the definition and using cor.

```
> cov(x,y) / (sd(x)*sd(y))
[1] 0.3551
> cor(x,y)
[1] 0.3551
```

Although the correlation is small, there is nevertheless a physical explanation for the correlation because both products are a result of incomplete combustion. A correlation of 0.36 typically corresponds to a slight visual impression that y tends to increase as x increases, although the points will be well scattered.

2.2.2 The ensemble and stationarity

The mean function of a time series model is

$$\mu(t) = E\left(x_t\right) \qquad (2.6)$$

and, in general, is a function of t. The expectation in this definition is an average taken across the *ensemble* of all the possible time series that might

---

[4] In statistics, *asymptotically* means as the sample size approaches infinity.

have been produced by the time series model (Fig. 2.2). The ensemble constitutes the entire population. If we have a time series model, we can simulate
more than one time series (see Chapter 4). However, with historical data, we
usually only have a single time series so all we can do, without assuming a
mathematical structure for the trend, is to estimate the mean at each sample
point by the corresponding observed value. In practice, we make estimates of
any apparent trend and seasonal effects in our data and remove them, using
decompose for example, to obtain time series of the random component. Then
time series models with a constant mean will be appropriate.

If the mean function is constant, we say that the time series model is
stationary in the mean. The sample estimate of the population mean, μ, is
the sample mean, $\bar{x}$:

$$\bar{x} = \sum_{t=1}^{n} x_t/n \tag{2.7}$$

Equation (2.7) does rely on an assumption that a sufficiently long time series
characterises the hypothetical model. Such models are known as *ergodic*, and
the models in this book are all ergodic.

2.2.3 Ergodic series*

A time series model that is stationary in the mean is *ergodic* in the mean if
the time average for a single time series tends to the ensemble mean as the
length of the time series increases:

$$\lim_{n \to \infty} \frac{\sum x_t}{n} = \mu \tag{2.8}$$

This implies that the time average is independent of the starting point. Given
that we usually only have a single time series, you may wonder how a time
series model can fail to be ergodic, or why we should want a model that is
not ergodic. Environmental and economic time series are single realisations of
a hypothetical time series model, and we simply define the underlying model
as ergodic.

There are, however, cases in which we can have many time series arising
from the same time series model. Suppose we investigate the acceleration at
the pilot seat of a new design of microlight aircraft in simulated random gusts
in a wind tunnel. Even if we have built two prototypes to the same design,
we cannot be certain they will have the same average acceleration response
because of slight differences in manufacture. In such cases, the number of time
series is equal to the number of prototypes. Another example is an experiment
investigating turbulent flows in some complex system. It is possible that we
will obtain qualitatively different results from different runs because they do
depend on initial conditions. It would seem better to run an experiment involving turbulence many times than to run it once for a much longer time.
The number of runs is the number of time series. It is straightforward to adapt

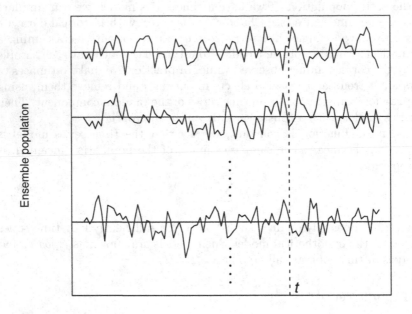

Time

Fig. 2.2. An ensemble of time series. The expected value $E(x_t)$ at a particular time t is the average taken over the entire population.

a stationary time series model to be non-ergodic by defining the means for the individual time series to be from some probability distribution.

2.2.4 Variance function

The variance function of a time series model that is stationary in the mean is

$$\sigma^2(t) = E\left[(x_t - \mu)^2\right] \tag{2.9}$$

which can, in principle, take a different value at every time t. But we cannot estimate a different variance at each time point from a single time series. To progress, we must make some simplifying assumption. If we assume the model is stationary in the variance, this constant population variance, σ^2, can be estimated from the sample variance:

$$\text{Var}(x) = \frac{\sum (x_t - \overline{x})^2}{n - 1} \tag{2.10}$$

In a time series analysis, sequential observations may be correlated. If the correlation is positive, $\text{Var}(x)$ will tend to underestimate the population variance in a short series because successive observations tend to be relatively similar. In most cases, this does not present a problem since the bias decreases rapidly as the length n of the series increases.

2.2.5 Autocorrelation

The mean and variance play an important role in the study of statistical distributions because they summarise two key distributional properties – a central location and the spread. Similarly, in the study of time series models, a key role is played by the *second-order properties*, which include the mean, variance, and serial correlation (described below).

Consider a time series model that is stationary in the mean and the variance. The variables may be correlated, and the model is *second-order stationary* if the correlation between variables depends only on the number of time steps separating them. The number of time steps between the variables is known as the *lag*. A correlation of a variable with itself at different times is known as *autocorrelation* or *serial correlation*. If a time series model is second-order stationary, we can define an *autocovariance function* (*acvf*), γ_k, as a function of the lag k:

$$\gamma_k = E\left[(x_t - \mu)(x_{t+k} - \mu)\right] \tag{2.11}$$

The function γ_k does not depend on t because the expectation, which is across the ensemble, is the same at all times t. This definition follows naturally from Equation (2.1) by replacing x with x_t and y with x_{t+k} and noting that the mean μ is the mean of both x_t and x_{t+k}. The lag k autocorrelation function (*acf*), ρ_k, is defined by

$$\rho_k = \frac{\gamma_k}{\sigma^2} \tag{2.12}$$

It follows from the definition that ρ_0 is 1.

It is possible to set up a second-order stationary time series model that has skewness; for example, one that depends on time t. Applications for such models are rare, and it is customary to drop the term 'second-order' and use 'stationary' on its own for a time series model that is at least second-order stationary. The term *strictly stationary* is reserved for more rigorous conditions.

The acvf and acf can be estimated from a time series by their sample equivalents. The sample acvf, c_k, is calculated as

$$c_k = \frac{1}{n}\sum_{t=1}^{n-k}\left(x_t - \bar{x}\right)\left(x_{t+k} - \bar{x}\right) \tag{2.13}$$

Note that the autocovariance at lag 0, c_0, is the variance calculated with a denominator n. Also, a denominator n is used when calculating c_k, although only $n - k$ terms are added to form the numerator. Adopting this definition constrains all sample autocorrelations to lie between -1 and 1. The sample acf is defined as

$$r_k = \frac{c_k}{c_0} \qquad (2.14)$$

We will demonstrate the calculations in R using a time series of wave heights (mm relative to still water level) measured at the centre of a wave tank. The sampling interval is 0.1 second and the record length is 39.7 seconds. The waves were generated by a wave maker driven by a pseudo-random signal that was programmed to emulate a rough sea. There is no trend and no seasonal period, so it is reasonable to suppose the time series is a realisation of a stationary process.

```
> www <- "http://www.massey.ac.nz/~pscowper/ts/wave.dat"
> wave.dat <- read.table (www, header=T) ; attach(wave.dat)
> plot(ts(waveht)) ; plot(ts(waveht[1:60]))
```

The upper plot in Figure 2.3 shows the entire time series. There are no outlying values. The lower plot is of the first sixty wave heights. We can see that there is a tendency for consecutive values to be relatively similar and that the form is like a rough sea, with a quasi-periodicity but no fixed frequency.

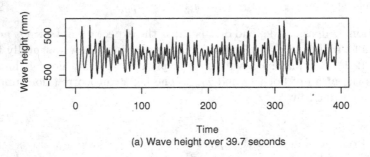

(a) Wave height over 39.7 seconds

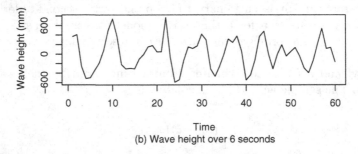

(b) Wave height over 6 seconds

Fig. 2.3. Wave height at centre of tank sampled at 0.1 second intervals.

The autocorrelations of x are stored in the vector acf(x)$acf, with the lag k autocorrelation located in acf(x)$acf[k+1]. For example, the lag 1 autocorrelation for waveht is

```
> acf(waveht)$acf[2]
```

```
[1] 0.47
```

The first entry, acf(waveht)$acf[1], is r_0 and equals 1. A scatter plot, such as Figure 2.1 for the Herald Square data, complements the calculation of the correlation and alerts us to any non-linear patterns. In a similar way, we can draw a scatter plot corresponding to each autocorrelation. For example, for lag 1 we plot(waveht[1:396],waveht[2:397]) to obtain Figure 2.4. Autocovariances are obtained by adding an argument to acf. The lag 1 autocovariance is given by

```
> acf(waveht, type = c("covariance"))$acf[2]
```

```
[1] 33328
```

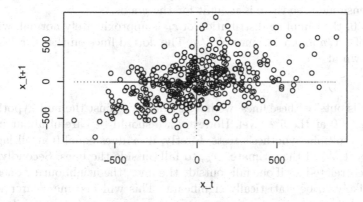

Fig. 2.4. Wave height pairs separated by a lag of 1.

2.3 The correlogram

2.3.1 General discussion

By default, the acf function produces a plot of r_k against k, which is called the *correlogram*. For example, Figure 2.5 gives the correlogram for the wave heights obtained from acf(waveht). In general, correlograms have the following features:

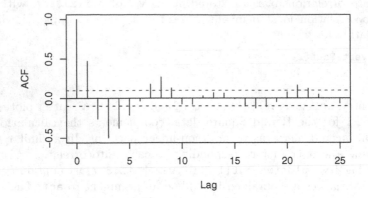

Fig. 2.5. Correlogram of wave heights.

- The x-axis gives the lag (k) and the y-axis gives the autocorrelation (r_k) at each lag. The unit of lag is the sampling interval, 0.1 second. Correlation is dimensionless, so there is no unit for the y-axis.
- If $\rho_k = 0$, the sampling distribution of r_k is approximately normal, with a mean of $-1/n$ and a variance of $1/n$. The dotted lines on the correlogram are drawn at

$$-\frac{1}{n} \pm \frac{2}{\sqrt{n}}$$

  If r_k falls outside these lines, we have evidence against the null hypothesis that $\rho_k = 0$ at the 5% level. However, we should be careful about interpreting multiple hypothesis tests. Firstly, if ρ_k does equal 0 at all lags k, we expect 5% of the estimates, r_k, to fall outside the lines. Secondly, the r_k are correlated, so if one falls outside the lines, the neighbouring ones are more likely to be statistically significant. This will become clearer when we simulate time series in Chapter 4. In the meantime, it is worth looking for statistically significant values at specific lags that have some practical meaning (for example, the lag that corresponds to the seasonal period, when there is one). For monthly series, a significant autocorrelation at lag 12 might indicate that the seasonal adjustment is not adequate.
- The lag 0 autocorrelation is always 1 and is shown on the plot. Its inclusion helps us compare values of the other autocorrelations relative to the theoretical maximum of 1. This is useful because, if we have a long time series, small values of r_k that are of no practical consequence may be statistically significant. However, some discernment is required to decide what constitutes a noteworthy autocorrelation from a practical viewpoint. Squaring the autocorrelation can help, as this gives the percentage of variability explained by a linear relationship between the variables. For example, a lag 1 autocorrelation of 0.1 implies that a linear dependency of x_t on x_{t-1}

would only explain 1% of the variability of x_t. It is a common fallacy to treat a statistically significant result as important when it has almost no practical consequence.

- The correlogram for wave heights has a well-defined shape that appears like a sampled damped cosine function. This is typical of correlograms of time series generated by an autoregressive model of order 2. We cover autoregressive models in Chapter 4.

If you look back at the plot of the air passenger bookings, there is a clear seasonal pattern and an increasing trend (Fig. 1.1). It is not reasonable to claim the time series is a realisation of a stationary model. But, whilst the population acf was defined only for a stationary time series model, the sample acf can be calculated for any time series, including deterministic signals. Some results for deterministic signals are helpful for explaining patterns in the acf of time series that we do not consider realisations of some stationary process:

- If you construct a time series that consists of a trend only, the integers from 1 up to 1000 for example, the acf decreases slowly and almost linearly from 1.
- If you take a large number of cycles of a discrete sinusoidal wave of any amplitude and phase, the acf is a discrete cosine function of the same period.
- If you construct a time series that consists of an arbitrary sequence of p numbers repeated many times, the correlogram has a dominant spike of almost 1 at lag p.

Usually a trend in the data will show in the correlogram as a slow decay in the autocorrelations, which are large and positive due to similar values in the series occurring close together in time. This can be seen in the correlogram for the air passenger bookings acf(AirPassengers) (Fig. 2.6). If there is seasonal variation, seasonal spikes will be superimposed on this pattern. The annual cycle appears in the air passenger correlogram as a cycle of the same period superimposed on the gradually decaying ordinates of the acf. This gives a maximum at a lag of 1 year, reflecting a positive linear relationship between pairs of variables (x_t, x_{t+12}) separated by 12-month periods. Conversely, because the seasonal trend is approximately sinusoidal, values separated by a period of 6 months will tend to have a negative relationship. For example, higher values tend to occur in the summer months followed by lower values in the winter months. A dip in the acf therefore occurs at lag 6 months (or 0.5 years). Although this is typical for seasonal variation that is approximated by a sinusoidal curve, other series may have patterns, such as high sales at Christmas, that contribute a single spike to the correlogram.

2.3.2 Example based on air passenger series

Although we want to know about trends and seasonal patterns in a time series, we do not necessarily rely on the correlogram to identify them. The main use

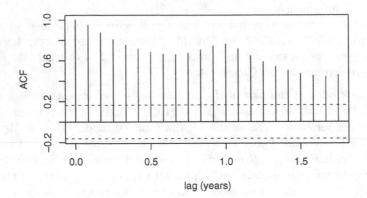

Fig. 2.6. Correlogram for the air passenger bookings over the period 1949–1960. The gradual decay is typical of a time series containing a trend. The peak at 1 year indicates seasonal variation.

of the correlogram is to detect autocorrelations in the time series after we have removed an estimate of the trend and seasonal variation. In the code below, the air passenger series is seasonally adjusted and the trend removed using `decompose`. To plot the random component and draw the correlogram, we need to remember that a consequence of using a centred moving average of 12 months to smooth the time series, and thereby estimate the trend, is that the first six and last six terms in the random component cannot be calculated and are thus stored in R as `NA`. The random component and correlogram are shown in Figures 2.7 and 2.8, respectively.

```
> data(AirPassengers)
> AP <- AirPassengers
> AP.decom <- decompose(AP, "multiplicative")
> plot(ts(AP.decom$random[7:138]))
> acf(AP.decom$random[7:138])
```

The correlogram in Figure 2.8 suggests either a damped cosine shape that is characteristic of an autoregressive model of order 2 (Chapter 4) or that the seasonal adjustment has not been entirely effective. The latter explanation is unlikely because the decomposition does estimate twelve independent monthly indices. If we investigate further, we see that the standard deviation of the original series from July until June is 109, the standard deviation of the series after subtracting the trend estimate is 41, and the standard deviation after seasonal adjustment is just 0.03.

```
> sd(AP[7:138])
```

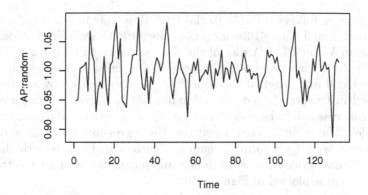

Fig. 2.7. The random component of the air passenger series after removing the trend and the seasonal variation.

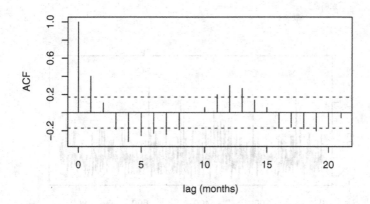

Fig. 2.8. Correlogram for the random component of air passenger bookings over the period 1949–1960.

```
[1] 109

> sd(AP[7:138] - AP.decom$trend[7:138])

[1] 41.1

> sd(AP.decom$random[7:138])

[1] 0.0335
```

The reduction in the standard deviation shows that the seasonal adjustment has been very effective.

2.3.3 Example based on the Font Reservoir series

Monthly effective inflows (m^3s^{-1}) to the Font Reservoir in Northumberland
for the period from January 1909 until December 1980 have been provided by
Northumbrian Water PLC. A plot of the data is shown in Figure 2.9. There
was a slight decreasing trend over this period, and substantial seasonal variation. The trend and seasonal variation have been estimated by regression,
as described in Chapter 5, and the residual series (`adflow`), which we analyse here, can reasonably be considered a realisation from a stationary time
series model. The main difference between the regression approach and using `decompose` is that the former assumes a linear trend, whereas the latter
smooths the time series without assuming any particular form for the trend.
The correlogram is plotted in Figure 2.10.

```
> www <- "http://www.massey.ac.nz/~pscowper/ts/Fontdsdt.dat"
> Fontdsdt.dat <- read.table(www, header=T)
> attach(Fontdsdt.dat)
> plot(ts(adflow), ylab = 'adflow')
> acf(adflow, xlab = 'lag (months)', main="")
```

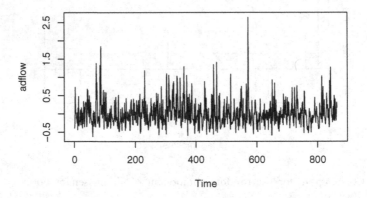

Fig. 2.9. Adjusted inflows to the Font Reservoir, 1909–1980.

There is a statistically significant correlation at lag 1. The physical interpretation is that the inflow next month is more likely than not to be above
average if the inflow this month is above average. Similarly, if the inflow this
month is below average it is more likely than not that next month's inflow
will be below average. The explanation is that the groundwater supply can be
thought of as a slowly discharging reservoir. If groundwater is high one month
it will augment inflows, and is likely to do so next month as well. Given this

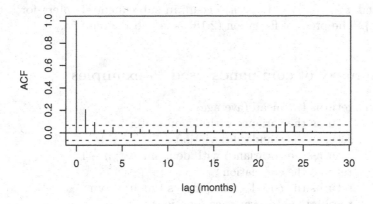

Fig. 2.10. Correlogram for adjusted inflows to the Font Reservoir, 1909–1980.

explanation, you may be surprised that the lag 1 correlation is not higher. The explanation for this is that most of the inflow is runoff following rainfall, and in Northumberland there is little correlation between seasonally adjusted rainfall in consecutive months. An exponential decay in the correlogram is typical of a first-order autoregressive model (Chapter 4). The correlogram of the adjusted inflows is consistent with an exponential decay. However, given the sampling errors for a time series of this length, estimates of autocorrelation at higher lags are unlikely to be statistically significant. This is not a practical limitation because such low correlations are inconsequential. When we come to identify suitable models, we should remember that there is no one correct model and that there will often be a choice of suitable models. We may make use of a specific statistical criterion such as *Akaike's information criterion*, introduced in Chapter 5, to choose a model, but this does not imply that the model is correct.

2.4 Covariance of sums of random variables

In subsequent chapters, second-order properties for several time series models are derived using the result shown in Equation (2.15). Let $x_1, x_2, \ldots, x_n$ and $y_1, y_2, \ldots, y_m$ be random variables. Then

$$\text{Cov}\left(\sum_{i=1}^{n} x_i, \sum_{j=1}^{m} y_j\right) = \sum_{i=1}^{n}\sum_{j=1}^{m} \text{Cov}(x_i, y_j) \qquad (2.15)$$

where $\text{Cov}(x, y)$ is the covariance between a pair of random variables x and y. The result tells us that the covariance of two sums of variables is the sum of all possible covariance pairs of the variables. Note that the special case of $n = m$ and $x_i = y_i$ $(i = 1, \ldots, n)$ occurs in subsequent chapters for a time series $\{x_t\}$. The proof of Equation (2.15) is left to Exercise 5a.

2.5 Summary of commands used in examples

mean	returns the mean (average)
var	returns the variance with denominator $n - 1$
sd	returns the standard deviation
cov	returns the covariance with denominator $n - 1$
cor	returns the correlation
acf	returns the correlogram (or sets the argument to obtain autocovariance function)

2.6 Exercises

1. On the book's website, you will find two small bivariate data sets that are not time series. Draw a scatter plot for each set and then calculate the correlation. Comment on your results.

   a) The data in the file varnish.dat are the amount of catalyst in a varnish, x, and the drying time of a set volume in a petri dish, y.

   b) The data in the file guesswhat.dat are data pairs. Can you see a pattern? Can you guess what they represent?

2. The following data are the volumes, relative to nominal contents of 750 ml, of 16 bottles taken consecutively from the filling machine at the Serendipity Shiraz vineyard:

   $$39, 35, 16, 18, 7, 22, 13, 18, 20, 9, -12, -11, -19, -9, -2, 16.$$

   The following are the volumes, relative to nominal contents of 750 ml, of consecutive bottles taken from the filling machine at the Cagey Chardonnay vineyard:

   $$47, -26, 42, -10, 27, -8, 16, 6, -1, 25, 11, 1, 25, 7, -5, 3$$

   The data are also available from the website in the file ch2ex2.dat.
   a) Produce time plots of the two time series.
   b) For each time series, draw a lag 1 scatter plot.
   c) Produce the acf for both time series and comment.

3. Carry out the following exploratory time series analysis using the global temperature series from §1.4.5.

   a) Decompose the series into the components trend, seasonal effect, and residuals. Plot these components. Would you expect these data to have a substantial seasonal component? Compare the standard deviation of the original series with the deseasonalised series. Produce a plot of the trend with a superimposed seasonal effect.

   b) Plot the correlogram of the residuals (random component) from part (a). Comment on the plot, with particular reference to any statistically significant correlations.

4. The monthly effective inflows (m^3s^{-1}) to the Font Reservoir are in the file Font.dat. Use decompose on the time series and then plot the correlogram of the random component. Compare this with Figure 2.10 and comment.

5. a) Prove Equation (2.15), using the following properties of summation, expectation, and covariance:

$$\sum_{i=1}^{n} x_i \sum_{j=1}^{m} y_j = \sum_{i=1}^{n} \sum_{j=1}^{m} x_i y_j$$
$$E\left[\sum_{i=1}^{n} x_i\right] = \sum_{i=1}^{n} E(x_i)$$
$$\text{Cov}(x, y) = E(xy) - E(x)E(y)$$

   b) By taking $n = m = 2$ and $x_i = y_i$ in Equation (2.15), derive the well-known result

$$\text{Var}(x + y) = \text{Var}(x) + \text{Var}(y) + 2\,\text{Cov}(x, y)$$

   c) Verify the result in part (b) above using R with x and y (CO and Benzoa, respectively) taken from §2.2.1.

3

Forecasting Strategies

3.1 Purpose

Businesses rely on forecasts of sales to plan production, justify marketing decisions, and guide research. A very efficient method of forecasting one variable is to find a related variable that leads it by one or more time intervals. The closer the relationship and the longer the lead time, the better this strategy becomes. The trick is to find a suitable lead variable. An Australian example is the Building Approvals time series published by the Australian Bureau of Statistics. This provides valuable information on the likely demand over the next few months for all sectors of the building industry. A variation on the strategy of seeking a leading variable is to find a variable that is associated with the variable we need to forecast and easier to predict.

In many applications, we cannot rely on finding a suitable leading variable and have to try other methods. A second approach, common in marketing, is to use information about the sales of similar products in the past. The influential Bass diffusion model is based on this principle. A third strategy is to make extrapolations based on present trends continuing and to implement adaptive estimates of these trends. The statistical technicalities of forecasting are covered throughout the book, and the purpose of this chapter is to introduce the general strategies that are available.

3.2 Leading variables and associated variables

3.2.1 Marine coatings

A leading international marine paint company uses statistics available in the public domain to forecast the numbers, types, and sizes of ships to be built over the next three years. One source of such information is *World Shipyard Monitor*, which gives brief details of orders in over 300 shipyards. The paint company has set up a database of ship types and sizes from which it can

P.S.P. Cowpertwait and A.V. Metcalfe, *Introductory Time Series with R*,
Use R, DOI 10.1007/978-0-387-88698-5_3,
© Springer Science+Business Media, LLC 2009

forecast the areas to be painted and hence the likely demand for paint. The company monitors its market share closely and uses the forecasts for planning production and setting prices.

3.2.2 Building approvals publication

Building approvals and building activity time series

The Australian Bureau of Statistics publishes detailed data on building approvals for each month, and, a few weeks later, the Building Activity Publication lists the value of building work done in each quarter. The data in the file ApprovActiv.dat are the total dwellings approved per month, averaged over the past three months, labelled "Approvals", and the value of work done over the past three months (chain volume measured in millions of Australian dollars at the reference year 2004–05 prices), labelled "Activity", from March 1996 until September 2006. We start by reading the data into R and then construct time series objects and plot the two series on the same graph using ts.plot (Fig. 3.1).

```
> www <- "http://www.massey.ac.nz/~pscowper/ts/ApprovActiv.dat"
> Build.dat <- read.table(www, header=T) ; attach(Build.dat)
> App.ts <- ts(Approvals, start = c(1996,1), freq=4)
> Act.ts <- ts(Activity, start = c(1996,1), freq=4)
> ts.plot(App.ts, Act.ts, lty = c(1,3))
```

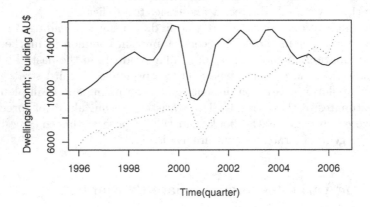

Fig. 3.1. Building approvals (solid line) and building activity (dotted line).

In Figure 3.1, we can see that the building activity tends to lag one quarter behind the building approvals, or equivalently that the building approvals appear to lead the building activity by a quarter. The *cross-correlation function*, which is abbreviated to *ccf*, can be used to quantify this relationship. A plot of the cross-correlation function against lag is referred to as a *cross-correlogram*.

Cross-correlation

Suppose we have time series models for variables x and y that are stationary in the mean and the variance. The variables may each be serially correlated, and correlated with each other at different time lags. The combined model is *second-order stationary* if all these correlations depend only on the lag, and then we can define the *cross covariance function* (*ccvf*), $\gamma_k(x, y)$, as a function of the lag, k:

$$\gamma_k(x, y) = E\left[(x_{t+k} - \mu_x)(y_t - \mu_y)\right] \tag{3.1}$$

This is not a symmetric relationship, and the variable x is lagging variable y by k. If x is the input to some physical system and y is the response, the cause will precede the effect, y will lag x, the ccvf will be 0 for positive k, and there will be spikes in the ccvf at negative lags. Some textbooks define ccvf with the variable y lagging when k is positive, but we have used the definition that is consistent with R. Whichever way you choose to define the ccvf,

$$\gamma_k(x, y) = \gamma_{-k}(y, x) \tag{3.2}$$

When we have several variables and wish to refer to the acvf of one rather than the ccvf of a pair, we can write it as, for example, $\gamma_k(x, x)$. The lag k cross-correlation function (*ccf*), $\rho_k(x, y)$, is defined by

$$\rho_k(x, y) = \frac{\gamma_k(x, y)}{\sigma_x \sigma_y}. \tag{3.3}$$

The ccvf and ccf can be estimated from a time series by their sample equivalents. The sample ccvf, $c_k(x, y)$, is calculated as

$$c_k(x, y) = \frac{1}{n} \sum_{t=1}^{n-k} (x_{t+k} - \bar{x})(y_t - \bar{y}) \tag{3.4}$$

The sample acf is defined as

$$r_k(x, y) = \frac{c_k(x, y)}{\sqrt{c_0(x, x)c_0(y, y)}} \tag{3.5}$$

Cross-correlation between building approvals and activity

The `ts.union` function binds time series with a common frequency, padding with 'NA's to the union of their time coverages. If `ts.union` is used within the `acf` command, R returns the correlograms for the two variables and the cross-correlograms in a single figure.

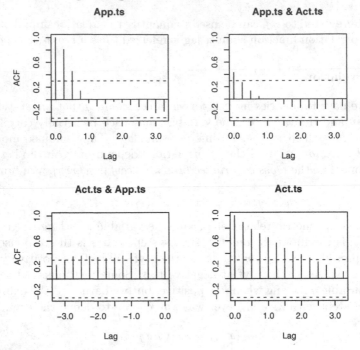

Fig. 3.2. Correlogram and cross-correlogram for building approvals and building activity.

```
> acf(ts.union(App.ts, Act.ts))
```

In Figure 3.2, the acfs for x and y are in the upper left and lower right frames, respectively, and the ccfs are in the lower left and upper right frames. The time unit for lag is one year, so a correlation at a lag of one quarter appears at 0.25. If the variables are independent, we would expect 5% of sample correlations to lie outside the dashed lines. Several of the cross-correlations at negative lags do pass these lines, indicating that the approvals time series is leading the activity. Numerical values can be printed using the print() function, and are 0.432, 0.494, 0.499, and 0.458 at lags of 0, 1, 2, and 3, respectively. The ccf can be calculated for any two time series that overlap, but if they both have trends or similar seasonal effects, these will dominate (Exercise 1). It may be that common trends and seasonal effects are precisely what we are looking for, but the population ccf is defined for stationary random processes and it is usual to remove the trend and seasonal effects before investigating cross-correlations. Here we remove the trend using decompose, which uses a centred moving average of the four quarters (see Fig. 3.3). We will discuss the use of ccf in later chapters.

```
> app.ran <- decompose(App.ts)$random
> app.ran.ts <- window (app.ran, start = c(1996, 3) )
> act.ran <- decompose (Act.ts)$random
> act.ran.ts <- window (act.ran, start = c(1996, 3) )
> acf (ts.union(app.ran.ts, act.ran.ts))
> ccf (app.ran.ts, act.ran.ts)
```

We again use print() to obtain the following table.

```
> print(acf(ts.union(app.ran.ts, act.ran.ts)))

      app.ran.ts        act.ran.ts
  1.000 ( 0.00)   0.123 ( 0.00)
  0.422 ( 0.25)   0.704 (-0.25)
 -0.328 ( 0.50)   0.510 (-0.50)
 -0.461 ( 0.75)  -0.135 (-0.75)
 -0.400 ( 1.00)  -0.341 (-1.00)
 -0.193 ( 1.25)  -0.187 (-1.25)
 ...

      app.ran.ts        act.ran.ts
  0.123 ( 0.00)   1.000 ( 0.00)
 -0.400 ( 0.25)   0.258 ( 0.25)
 -0.410 ( 0.50)  -0.410 ( 0.50)
 -0.250 ( 0.75)  -0.411 ( 0.75)
  0.071 ( 1.00)  -0.112 ( 1.00)
  0.353 ( 1.25)   0.180 ( 1.25)
 ...
```

The ccf function produces a single plot, shown in Figure 3.4, and again shows the lagged relationship. The Australian Bureau of Statistics publishes the building approvals by state and by other categories, and specific sectors of the building industry may find higher correlations between demand for their products and one of these series than we have seen here.

3.2.3 Gas supply

Gas suppliers typically have to place orders for gas from offshore fields 24 hours ahead. Variation about the average use of gas, for the time of year, depends on temperature and, to some extent, humidity and wind speed. Coleman et al. (2001) found that the weather accounts for 90% of this variation in the United Kingdom. Weather forecasts for the next 24 hours are now quite accurate and are incorporated into the forecasting procedure.

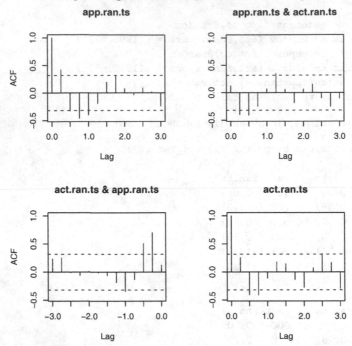

Fig. 3.3. Correlogram and cross-correlogram of the random components of building approvals and building activity after using `decompose`.

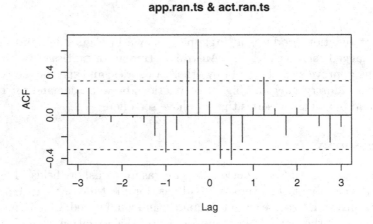

Fig. 3.4. Cross-correlogram of the random components of building approvals and building activity after using `decompose`.

3.3 Bass model

3.3.1 Background

Frank Bass published a paper describing his mathematical model, which quantified the theory of adoption and diffusion of a new product by society (Rogers, 1962), in *Management Science* nearly fifty years ago (Bass, 1969). The mathematics is straightforward, and the model has been influential in marketing. An entrepreneur with a new invention will often use the Bass model when making a case for funding. There is an associated demand for market research, as demonstrated, for example, by the Marketing Science Centre at the University of South Australia becoming the Ehrenberg-Bass Institute for Marketing Science in 2005.

3.3.2 Model definition

The Bass formula for the number of people, N_t, who have bought a product at time t depends on three parameters: the total number of people who eventually buy the product, m; the *coefficient of innovation*, p; and the *coefficient of imitation*, q. The Bass formula is

$$N_{t+1} = N_t + p(m - N_t) + qN_t(m - N_t)/m \qquad (3.6)$$

According to the model, the increase in sales, $N_{t+1} - N_t$, over the next time period is equal to the sum of a fixed proportion p and a time varying proportion $q\frac{N_t}{m}$ of people who will eventually buy the product but have not yet done so. The rationale for the model is that initial sales will be to people who are interested in the novelty of the product, whereas later sales will be to people who are drawn to the product after seeing their friends and acquaintances use it. Equation (3.6) is a difference equation and its solution is

$$N_t = m\frac{1 - e^{-(p+q)t}}{1 + (q/p)e^{-(p+q)t}} \qquad (3.7)$$

It is easier to verify this result for the continuous-time version of the model.

3.3.3 Interpretation of the Bass model*

One interpretation of the Bass model is that the time from product launch until purchase is assumed to have a probability distribution that can be parametrised in terms of p and q. A plot of sales per time unit against time is obtained by multiplying the probability density by the number of people, m, who eventually buy the product. Let $f(t)$, $F(t)$, and $h(t)$ be the density, cumulative distribution function (cdf), and hazard, respectively, of the distribution of time until purchase. The definition of the hazard is

$$h(t) = \frac{f(t)}{1 - F(t)} \tag{3.8}$$

The interpretation of the hazard is that if it is multiplied by a small time increment it gives the probability that a random purchaser who has not yet made the purchase will do so in the next small time increment (Exercise 2). Then the continuous time model of the Bass formula can be expressed in terms of the hazard:

$$h(t) = p + qF(t) \tag{3.9}$$

Equation (3.6) is the discrete form of Equation (3.9) (Exercise 2). The solution of Equation (3.8), with $h(t)$ given by Equation (3.9), for $F(t)$ is

$$F(t) = \frac{1 - e^{-(p+q)t}}{1 + (q/p)e^{-(p+q)t}} \tag{3.10}$$

Two special cases of the distribution are the *exponential distribution* and *logistic distribution*, which arise when $q = 0$ and $p = 0$, respectively. The logistic distribution closely resembles the normal distribution (Exercise 3). Cumulative sales are given by the product of m and $F(t)$. The pdf is the derivative of Equation (3.10):

$$f(t) = \frac{(p+q)^2 e^{-(p+q)t}}{p\left[1 + (q/p)e^{-(p+q)t}\right]^2} \tag{3.11}$$

Sales per unit time at time t are

$$S(t) = mf(t) = \frac{m(p+q)^2 e^{-(p+q)t}}{p\left[1 + (q/p)e^{-(p+q)t}\right]^2} \tag{3.12}$$

The time to peak is

$$t_{\text{peak}} = \frac{\log(q) - \log(p)}{p + q} \tag{3.13}$$

3.3.4 Example

We show a typical Bass curve by fitting Equation (3.12) to yearly sales of VCRs in the US home market between 1980 and 1989 (Bass website) using the R non-linear least squares function `nls`. The variable `T79` is the year from 1979, and the variable `Tdelt` is the time from 1979 at a finer resolution of 0.1 year for plotting the Bass curves. The cumulative sum function `cumsum` is useful for monitoring changes in the mean level of the process (Exercise 8).

```
> T79 <- 1:10
> Tdelt <- (1:100) / 10
> Sales <- c(840,1470,2110,4000, 7590, 10950, 10530, 9470, 7790, 5890)
> Cusales <- cumsum(Sales)
> Bass.nls <- nls(Sales ~ M * ( ((P+Q)^2 / P) * exp(-(P+Q) * T79) ) /
  (1+(Q/P)*exp(-(P+Q)*T79))^2, start = list(M=60630, P=0.03, Q=0.38))
> summary(Bass.nls)
```

```
Parameters:
    Estimate Std. Error t value Pr(>|t|)
M 6.798e+04  3.128e+03    21.74 1.10e-07 ***
P 6.594e-03  1.430e-03     4.61  0.00245 **
Q 6.381e-01  4.140e-02    15.41 1.17e-06 ***

Residual standard error: 727.2 on 7 degrees of freedom
```

The final estimates for m, p, and q, rounded to two significant places, are 68000, 0.0066, and 0.64 respectively. The starting values for P and Q are p and q for a typical product. We assume the sales figures are prone to error and estimate the total sales, m, setting the starting value for M to the recorded total sales. The data and fitted curve can be plotted using the code below (see Fig. 3.5 and 3.6):

```
> Bcoef <- coef(Bass.nls)
> m <- Bcoef[1]
> p <- Bcoef[2]
> q <- Bcoef[3]
> ngete <- exp(-(p+q) * Tdelt)
> Bpdf <- m * ( (p+q)^2 / p ) * ngete / (1 + (q/p) * ngete)^2
> plot(Tdelt, Bpdf, xlab = "Year from 1979",
                    ylab = "Sales per year", type='l')
> points(T79, Sales)
> Bcdf <- m * (1 - ngete)/(1 + (q/p)*ngete)
> plot(Tdelt, Bcdf, xlab = "Year from 1979",
                    ylab = "Cumulative sales", type='l')
> points(T79, Cusales)
```

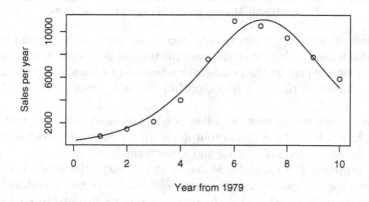

Fig. 3.5. Bass sales curve fitted to sales of VCRs in the US home market, 1980–1989.

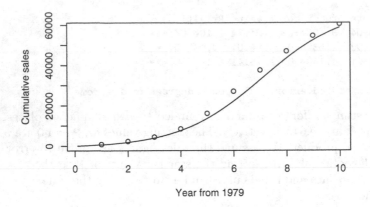

Fig. 3.6. Bass cumulative sales curve, obtained as the integral of the sales curve, and cumulative sales of VCRs in the US home market, 1980–1989.

It is easy to fit a curve to past sales data. The importance of the Bass curve in marketing is in forecasting, which needs values for the parameters m, p, and q. Plausible ranges for the parameter values can be based on published data for similar categories of past inventions, and a few examples follow.

Product	m	p	q	Reference
Typical product	-	0.030	0.380	VBM[1]
35 mm projectors, 1965–1986	3.37 million	0.009	0.173	Bass[2]
Overhead projectors, 1960–1970	0.961 million	0.028	0.311	Bass
PCs, 1981–2010	3.384 billion	0.001	0.195	Bass

[1]Value-Based Management; [2]Frank M. Bass, 1999.

Although the forecasts are inevitably uncertain, they are the best information available when making marketing and investment decisions. A prospectus for investors or a report to the management team will typically include a set of scenarios based on the most likely, optimistic, and pessimistic sets of parameters.

The basic Bass model does not allow for replacement sales and multiple purchases. Extensions of the model that allow for replacement sales, multiple purchases, and the effects of pricing and advertising in a competitive market have been proposed (for example, Mahajan et al. 2000). However, there are several reasons why these refinements may be of less interest to investors than you might expect. The first is that the profit margin on manufactured goods, such as innovative electronics and pharmaceuticals, will drop dramatically once patent protection expires and competitors enter the market. A second reason is that successful inventions are often superseded by new technology, as

VCRs have been by DVD players, and replacement sales are limited. Another reason is that many investors are primarily interested in a relatively quick return on their money. You are asked to consider Bass models for sales of two recent 3G mobile communication devices in Exercise 4.

3.4 Exponential smoothing & the Holt-Winters method

3.4.1 Exponential smoothing

Our objective is to predict some future value x_{n+k} given a past history $\{x_1, x_2, \ldots, x_n\}$ of observations up to time n. In this subsection we assume there is no systematic trend or seasonal effects in the process, or that these have been identified and removed. The mean of the process can change from one time step to the next, but we have no information about the likely direction of these changes. A typical application is forecasting sales of a well-established product in a stable market. The model is

$$x_t = \mu_t + w_t \qquad (3.14)$$

where μ_t is the non-stationary mean of the process at time t and w_t are independent random deviations with a mean of 0 and a standard deviation σ. We will follow the notation in R and let a_t be our estimate of μ_t. Given that there is no systematic trend, an intuitively reasonable estimate of the mean at time t is given by a weighted average of our observation at time t and our estimate of the mean at time $t - 1$:

$$a_t = \alpha x_t + (1 - \alpha)a_{t-1} \qquad 0 < \alpha < 1 \qquad (3.15)$$

The a_t in Equation (3.15) is the *exponentially weighted moving average (EWMA)* at time t. The value of α determines the amount of smoothing, and it is referred to as the *smoothing parameter*. If α is near 1, there is little smoothing and a_t is approximately x_t. This would only be appropriate if the changes in the mean level were expected to be large by comparison with σ. At the other extreme, a value of α near 0 gives highly smoothed estimates of the mean level and takes little account of the most recent observation. This would only be appropriate if the changes in the mean level were expected to be small compared with σ. A typical compromise figure for α is 0.2 since in practice we usually expect that the change in the mean between time $t - 1$ and time t is likely to be smaller than σ. Alternatively, R can provide an estimate for α, and we discuss this option below. Since we have assumed that there is no systematic trend and that there are no seasonal effects, forecasts made at time n for any lead time are just the estimated mean at time n. The forecasting equation is

$$\hat{x}_{n+k|n} = a_n \qquad k = 1, 2, \ldots \qquad (3.16)$$

Equation (3.15), for a_t, can be rewritten in two other useful ways. Firstly, we can write the sum of a_{t-1} and a proportion of the one-step-ahead forecast error, $x_t - a_{t-1}$,

$$a_t = \alpha(x_t - a_{t-1}) + a_{t-1} \qquad (3.17)$$

Secondly, by repeated back substitution we obtain

$$a_t = \alpha x_t + \alpha(1-\alpha)x_{t-1} + \alpha(1-\alpha)^2 x_{t-2} + \ldots \qquad (3.18)$$

When written in this form, we see that a_t is a linear combination of the current and past observations, with more weight given to the more recent observations. The restriction $0 < \alpha < 1$ ensures that the weights $\alpha(1-\alpha)^i$ become smaller as i increases. Note that these weights form a geometric series, and the sum of the infinite series is unity (Exercise 5). We can avoid the infinite regression by specifying $a_1 = x_1$ in Equation (3.15).

For any given α, the model in Equation (3.17) together with the starting value $a_1 = x_1$ can be used to calculate a_t for $t = 2, 3, \ldots, n$. One-step-ahead prediction errors, e_t, are given by

$$e_t = x_t - \hat{x}_{t|t-1} = x_t - a_{t-1} \qquad (3.19)$$

By default, R obtains a value for the smoothing parameter, α, by minimising the sum of squared one-step-ahead prediction errors ($SS1PE$):

$$SS1PE = \sum_{t=2}^{n} e_t^2 = e_2^2 + e_3^2 + \ldots + e_n^2 \qquad a_1 = x_1 \qquad (3.20)$$

However, calculating α in this way is not necessarily the best practice. If the time series is long and the mean has changed little, the value of α will be small. In the specific case where the mean of the process does not change, the optimum value for α is $\frac{1}{n}$. An exponential smoothing procedure set up with a small value of α will be slow to respond to any unexpected change in the market, as occurred in sales of videotapes, which plummeted after the invention of DVDs.

Complaints to a motoring organisation

The number of letters of complaint received each month by a motoring organisation over the four years 1996 to 1999 are available on the website. At the beginning of the year 2000, the organisation wishes to estimate the current level of complaints and investigate any trend in the level of complaints. We should first plot the data, and, even though there are only four years of data, we should check for any marked systematic trend or seasonal effects.

```
> www <- "http://www.massey.ac.nz/~pscowper/ts/motororg.dat"
> Motor.dat <- read.table(www, header = T); attach(Motor.dat)
> Comp.ts <- ts(complaints, start = c(1996, 1), freq = 12)
> plot(Comp.ts, xlab = "Time / months", ylab = "Complaints")
```

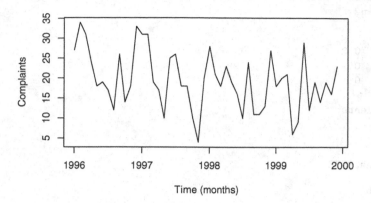

Fig. 3.7. Monthly numbers of letters of complaint received by a motoring organisation.

There is no evidence of a systematic trend or seasonal effects, so it seems
reasonable to use exponential smoothing for this time series. Exponential
smoothing is a special case of the Holt-Winters algorithm, which we introduce in the next section, and is implemented in R using the `HoltWinters`
function with the additional parameters set to 0. If we do not specify a value
for α, R will find the value that minimises the one-step-ahead prediction error.

```
> Comp.hw1 <- HoltWinters(complaints, beta = 0, gamma = 0) ; Comp.hw1
> plot(Comp.hw1)
```

```
Holt-Winters exponential smoothing without trend and without seasonal
component.

Smoothing parameters:
 alpha:  0.143
 beta :  0
 gamma:  0

Coefficients:
    [,1]
 a 17.70

> Comp.hw1$SSE

[1] 2502
```

The estimated value of the mean number of letters of complaint per month
at the end of 1999 is 17.7. The value of α that gives a minimum SS1PE, of
2502, is 0.143. We now compare these results with those obtained if we specify
a value for α of 0.2.

```
> Comp.hw2 <- HoltWinters(complaints, alpha = 0.2, beta=0, gamma=0)
> Comp.hw2

...
 alpha:  0.2
 beta :  0
 gamma:  0

Coefficients:
    [,1]
a 17.98

> Comp.hw2$SSE

[1] 2526
```

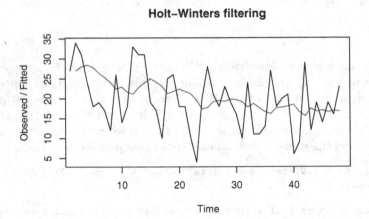

Fig. 3.8. Monthly numbers of letters and exponentially weighted moving average.

The estimated value of the mean number of letters of complaint per month at the end of 1999 is now 18.0, and the SS1PE has increased slightly to 2526. The advantage of letting R estimate a value for α is that it is optimum for a practically important criterion, SS1PE, and that it removes the need to make a choice. However, the optimum estimate can be close to 0 if we have a long time series over a stable period, and this makes the EWMA unresponsive to any future change in mean level. From Figure 3.8, it seems that there was a decrease in the number of complaints at the start of the period and a slight rise towards the end, although this has not yet affected the exponentially weighted moving average.

3.4.2 Holt-Winters method

We usually have more information about the market than exponential smoothing can take into account. Sales are often seasonal, and we may expect trends to be sustained for short periods at least. But trends will change. If we have a successful invention, sales will increase initially but then stabilise before declining as competitors enter the market. We will refer to the change in level from one time period to the next as the *slope*.[1] Seasonal patterns can also change due to vagaries of fashion and variation in climate, for example. The Holt-Winters method was suggested by Holt (1957) and Winters (1960), who were working in the School of Industrial Administration at Carnegie Institute of Technology, and uses exponentially weighted moving averages to update estimates of the seasonally adjusted mean (called the *level*), slope, and seasonals.

The Holt-Winters method generalises Equation (3.15), and the additive seasonal form of their updating equations for a series $\{x_t\}$ with period p is

$$
\left.
\begin{aligned}
a_t &= \alpha(x_t - s_{t-p}) + (1 - \alpha)(a_{t-1} + b_{t-1}) \\
b_t &= \beta(a_t - a_{t-1}) + (1 - \beta)b_{t-1} \\
s_t &= \gamma(x_t - a_t) + (1 - \gamma)s_{t-p}
\end{aligned}
\right\}
\qquad (3.21)
$$

where a_t, b_t, and s_t are the estimated level,[2] slope, and seasonal effect at time t, and α, β, and γ are the smoothing parameters. The first updating equation takes a weighted average of our latest observation, with our existing estimate of the appropriate seasonal effect subtracted, and our forecast of the level made one time step ago. The one-step-ahead forecast of the level is the sum of the estimates of the level and slope at the time of forecast. A typical choice of the weight α is 0.2. The second equation takes a weighted average of our previous estimate and latest estimate of the slope, which is the difference in the estimated level at time t and the estimated level at time $t - 1$. Note that the second equation can only be used after the first equation has been applied to get a_t. Finally, we have another estimate of the seasonal effect, from the difference between the observation and the estimate of the level, and we take a weighted average of this and the last estimate of the seasonal effect for this season, which was made at time $t - p$. Typical choices of the weights β and γ are 0.2. The updating equations can be started with $a_1 = x_1$ and initial slope, b_1, and seasonal effects, $s_1, \ldots, s_p$, reckoned from experience, estimated from the data in some way, or set at 0. The default in R is to use values obtained from the decompose procedure.

The forecasting equation for x_{n+k} made after the observation at time n is

$$
\hat{x}_{n+k|n} = a_n + kb_n + s_{n+k-p} \qquad\qquad k \le p \qquad (3.22)
$$

---

[1] When describing the Holt-Winters procedure, the R help and many textbooks refer to the slope as the trend.

[2] The mean of the process is the sum of the level and the appropriate seasonal effect.

where a_n is the estimated level and b_n is the estimated slope, so $a_n + k b_n$ is the expected level at time $n + k$ and s_{n+k-p} is the exponentially weighted estimate of the seasonal effect made at time $n = k - p$. For example, for monthly data ($p = 12$), if time $n + 1$ occurs in January, then s_{n+1-12} is the exponentially weighted estimate of the seasonal effect for January made in the previous year. The forecasting equation can be used for lead times between $(m-1)p+1$ and mp, but then the most recent exponentially weighted estimate of the seasonal effect available will be $s_{n+k-(m-1)p}$.

The Holt-Winters algorithm with multiplicative seasonals is

$$\left. \begin{array}{l} a_n = \alpha \left(\frac{x_n}{s_{n-p}} \right) + (1 - \alpha)(a_{n-1} + b_{n-1}) \\ b_n = \beta(a_n - a_{n-1}) + (1 - \beta)b_{n-1} \\ s_n = \gamma \left(\frac{x_n}{a_n} \right) + (1 - \gamma)s_{n-p} \end{array} \right\} \qquad (3.23)$$

The forecasting equation for x_{n+k} made after the observation at time n becomes

$$\hat{x}_{n+k|n} = (a_n + k b_n)s_{n+k-p} \qquad\qquad k \leq p \qquad (3.24)$$

In R, the function HoltWinters can be used to estimate smoothing parameters for the Holt-Winters model by minimising the one-step-ahead prediction errors (SS1PE).

Sales of Australian wine

The data in the file wine.dat are monthly sales of Australian wine by category, in thousands of litres, from January 1980 until July 1995. The categories are fortified white, dry white, sweet white, red, rose, and sparkling. The sweet white wine time series is plotted in Figure 3.9, and there is a dramatic increase in sales in the second half of the 1980s followed by a reduction to a level well above the starting values. The seasonal variation looks as though it would be better modelled as multiplicative, and comparison of the SS1PE for the fitted models confirms this (Exercise 6). Here we present results for the model with multiplicative seasonals only. The Holt-Winters components and fitted values are shown in Figures 3.10 and 3.11 respectively.

```
> www <- "http://www.massey.ac.nz/~pscowper/ts/wine.dat"
> wine.dat <- read.table(wine, header = T) ; attach (wine.dat)
> sweetw.ts <- ts(sweetw, start = c(1980,1), freq = 12)
> plot(sweetw.ts, xlab= "Time (months)", ylab = "sales (1000 litres)")
> sweetw.hw <- HoltWinters (sweetw.ts, seasonal = "mult")
> sweetw.hw ; sweetw.hw$coef ; sweetw.hw$SSE

...
Smoothing parameters:
 alpha:   0.4107
 beta :   0.0001516
```

```
gamma:   0.4695
...

> sqrt(sweetw.hw$SSE/length(sweetw))
[1] 50.04
> sd(sweetw)
[1] 121.4

> plot (sweetw.hw$fitted)
> plot (sweetw.hw)
```

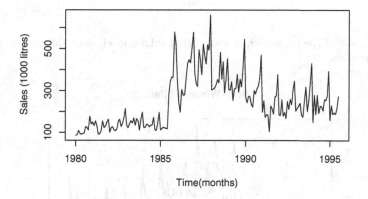

Fig. 3.9. Sales of Australian sweet white wine.

The optimum values for the smoothing parameters, based on minimising the one-step ahead prediction errors, are 0.4107, 0.0001516, and 0.4695 for α, β, and γ, respectively. It follows that the level and seasonal variation adapt rapidly whereas the trend is slow to do so. The coefficients are the estimated values of the level, slope, and multiplicative seasonals from January to December available at the latest time point ($t = n = 187$), and these are the values that will be used for predictions (Exercise 6). Finally, we have calculated the mean square one-step-ahead prediction error, which equals 50, and have compared it with the standard deviation of the original time series which is 121. The decrease is substantial, but a more testing comparison would be with the mean one-step-ahead prediction error if we forecast the next month's sales as equal to this month's sales (Exercise 6). Also, in Exercise 6 you are asked to investigate the performance of the Holt-Winters algorithm if the three smoothing parameters are all set equal to 0.2 and if the values for the parameters are optimised at each time step.

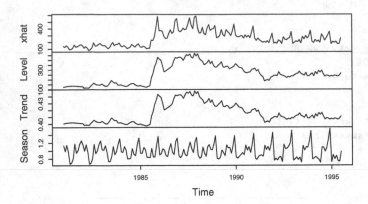

Fig. 3.10. Sales of Australian white wine: fitted values; level; slope (labelled trend); seasonal variation.

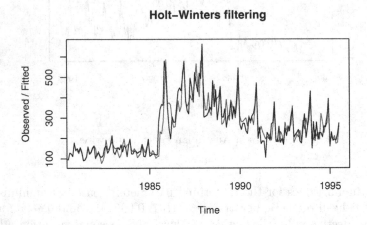

Fig. 3.11. Sales of Australian white wine and Holt-Winters fitted values.

3.4.3 Four-year-ahead forecasts for the air passenger data

The seasonal effect for the air passenger data of §1.4.1 appeared to increase with the trend, which suggests that a 'multiplicative' seasonal component be used in the Holt-Winters procedure. The Holt-Winters fit is impressive – see Figure 3.12. The `predict` function in R can be used with the fitted model to make forecasts into the future (Fig. 3.13).

```
> AP.hw <- HoltWinters(AP, seasonal = "mult")
> plot(AP.hw)
```

```
> AP.predict <- predict(AP.hw, n.ahead = 4 * 12)
> ts.plot(AP, AP.predict, lty = 1:2)
```

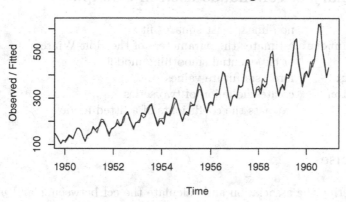

Fig. 3.12. Holt-Winters fit for air passenger data.

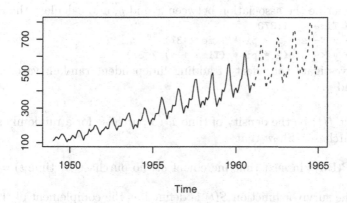

Fig. 3.13. Holt-Winters forecasts for air passenger data for 1961–1964 shown as dotted lines.

The estimates of the model parameters, which can be obtained from `AP.hw$alpha`, `AP.hw$beta`, and `AP.hw$gamma`, are $\hat{\alpha} = 0.274$, $\hat{\beta} = 0.0175$, and $\hat{\gamma} = 0.877$. It should be noted that the extrapolated forecasts are based entirely on the trends in the period during which the model was fitted and would be a sensible prediction assuming these trends continue. Whilst the extrapolation in Figure 3.12 looks visually appropriate, unforeseen events could lead to completely different future values than those shown here.

3.5 Summary of commands used in examples

`nls`	non-linear least squares fit
`HoltWinters`	estimates the parameters of the Holt-Winters or exponential smoothing model
`predict`	forecasts future values
`ts.union`	create the union of two series
`coef`	extracts the coefficients of a fitted model

3.6 Exercises

1. a) Describe the association and calculate the ccf between x and y for k equal to 1, 10, and 100.
    ```
    > w <- 1:100
    > x <- w + k * rnorm(100)
    > y <- w + k * rnorm(100)
    > ccf(x, y)
    ```
    b) Describe the association between x and y, and calculate the ccf.
    ```
    > Time <- 1:370
    > x <- sin(2 * pi * Time / 37)
    > y <- sin(2 * pi * (Time + 4) / 37)
    ```
    Investigate the effect of adding independent random variation to x and y.

2. a) Let $f(t)$ be the density of time T to purchase for a randomly selected purchaser. Show that

    $$P(\text{Buys in next time increment } \delta t| \text{ no purchase by time } t) = h(t)\delta t$$

    b) The survivor function $S(t)$ is defined as the complement of the cdf

    $$S(t) = 1 - F(t)$$

    Show that $S(t) = \exp(-\int_0^t h(u)\, du)$ and $E[T] = \int_0^\infty S(t)\, dt$.

    c) Explain how Equation (3.6) is the discrete form of Equation (3.9).

3. a) Verify that the solution of Equation (3.8), with $h(t)$ given by Equation (3.9), for $F(t)$ is Equation (3.10).

b) The logistic distribution has the cdf: $F(t) = \{1 + \exp(-(t-\mu)/b)\}^{-1}$, with mean μ and standard deviation $b\pi/\sqrt{3}$. Plot the cdf of the logistic distribution with a mean 0 and standard deviation 1 against the cdf of the standard normal distribution.

c) Show that the time to peak of the Bass curve is given by Equation (3.13). What does this reduce to for the exponential and logistic distributions?

4. The *Independent* on July 11, 2008 reported the launch of Apple's iPhone. A Deutsche Bank analyst predicted Apple would sell 10.5 million units during the year. The company was reported to have a target of 10 million units worldwide for 2008. Initial demand is predicted to exceed supply. Carphone Warehouse reportedly sold their online allocations within hours and expect to sell out at most of their UK shops. The report stated that there were 60,000 applications for 500 iPhones on the Hutchison Telecommunications website in Hong Kong.

   a) Why is a Bass model without replacement or multiple purchases likely to be realistic for this product?

   b) Suggest plausible values for the parameters p, q, and m for the model in (a), and give a likely range for these parameters. How does the shape of the cumulative sales curve vary with the parameter values?

   c) How could you allow for high initial sales with the Bass model?

5. a) Write the sum of n terms in a geometric progression with a first term a and a common ratio r as

$$S_n = a + ar + ar^2 + \ldots + ar^{n-1}$$

   Subtract rS_n from S_n and rearrange to obtain the formula for the sum of n terms:

$$S_n = \frac{a(1 - r^n)}{1 - r}$$

   b) Under what conditions does the sum of n terms of a geometric progression tend to a finite sum as n tends to infinity? What is this sum?

   c) Obtain an expression for the sum of the weights in an EWMA if we specify $a_1 = x_1$ in Equation (3.15).

   d) Suppose x_t happens to be a sequence of independent variables with a constant mean and a constant variance σ^2. What is the variance of a_t if we specify $a_1 = x_1$ in Equation (3.15)?

6. Refer to the sweet white wine sales (§3.4.2).

   a) Use the `HoltWinters` procedure with α, β and γ set to 0.2 and compare the SS1PE with the minimum obtained with R.

   b) Use the `HoltWinters` procedure on the logarithms of sales and compare SS1PE with that obtained using sales.

c) What is the SS1PE if you predict next month's sales will equal this month's sales?

d) This is rather harder: What is the SS1PE if you find the optimum α, β and γ from the data available at each time step before making the one-step-ahead prediction?

7. Continue the following exploratory time series analysis using the global temperature series from §1.4.5.

a) Produce a time plot of the data. Plot the aggregated annual mean series and a boxplot that summarises the observed values for each season, and comment on the plots.

b) Decompose the series into the components trend, seasonal effect, and residuals, and plot the decomposed series. Produce a plot of the trend with a superimposed seasonal effect.

c) Plot the correlogram of the residuals from question 7b. Comment on the plot, explaining any 'significant' correlations at significant lags.

d) Fit an appropriate Holt-Winters model to the monthly data. Explain why you chose that particular Holt-Winters model, and give the parameter estimates.

e) Using the fitted model, forecast values for the years 2005–2010. Add these forecasts to a time plot of the original series. Under what circumstances would these forecasts be valid? What comments of caution would you make to an economist or politician who wanted to use these forecasts to make statements about the potential impact of global warming on the world economy?

8. A cumulative sum plot is useful for monitoring changes in the mean of a process. If we have a time series composed of observations x_t at times t with a target value of τ, the CUSUM chart is a plot of the cumulative sums of the deviations from target, cs_t, against t. The formula for cs_t at time t is

$$cs_t = \sum_{i=1}^{t} (x_i - \tau)$$

The R function cumsum calculates a cumulative sum. Plot the CUSUM for the motoring organisation complaints with a target of 18.

9. Using the motor organisation complaints series, refit the exponential smoothing model with weights $\alpha = 0.01$ and $\alpha = 0.99$. In each case, extract the last residual from the fitted model and verify that the last residual satisfies Equation (3.19). Redraw Figure 3.8 using the new values of α, and comment on the plots, explaining the main differences.

4

Basic Stochastic Models

4.1 Purpose

So far, we have considered two approaches for modelling time series. The
first is based on an assumption that there is a fixed seasonal pattern about a
trend. We can estimate the trend by local averaging of the deseasonalised data,
and this is implemented by the R function decompose. The second approach
allows the seasonal variation and trend, described in terms of a level and slope,
to change over time and estimates these features by exponentially weighted
averages. We used the HoltWinters function to demonstrate this method.

When we fit mathematical models to time series data, we refer to the discrepancies between the fitted values, calculated from the model, and the data
as a *residual error series*. If our model encapsulates most of the deterministic
features of the time series, our residual error series should appear to be a realisation of independent random variables from some probability distribution.
However, we often find that there is some structure in the residual error series,
such as consecutive errors being positively correlated, which we can use to improve our forecasts and make our simulations more realistic. We assume that
our residual error series is stationary, and in Chapter 6 we introduce models
for stationary time series.

Since we judge a model to be a good fit if its residual error series appears
to be a realisation of independent random variables, it seems natural to build
models up from a model of independent random variation, known as discrete
white noise. The name 'white noise' was coined in an article on heat radiation
published in *Nature* in April 1922, where it was used to refer to series that
contained *all* frequencies in *equal* proportions, analogous to white light. The
term *purely random* is sometimes used for white noise series. In §4.3 we define a
fundamental non-stationary model based on discrete white noise that is called
the *random walk*. It is sometimes an adequate model for financial series and is
often used as a standard against which the performance of more complicated
models can be assessed.

P.S.P. Cowpertwait and A.V. Metcalfe, *Introductory Time Series with R*,
Use R, DOI 10.1007/978-0-387-88698-5_4,
© Springer Science+Business Media, LLC 2009

4.2 White noise

4.2.1 Introduction

A residual error is the difference between the observed value and the model predicted value at time t. If we suppose the model is defined for the variable y_t and $\hat{y}_t$ is the value predicted by the model, the residual error x_t is

$$x_t = y_t - \hat{y}_t \tag{4.1}$$

As the residual errors occur in time, they form a time series: $x_1, x_2, \ldots, x_n$.

In Chapter 2, we found that features of the historical series, such as the trend or seasonal variation, are reflected in the correlogram. Thus, if a model has accounted for all the serial correlation in the data, the residual series would be serially *uncorrelated*, so that a correlogram of the residual series would exhibit no obvious patterns. This ideal motivates the following definition.

4.2.2 Definition

A time series $\{w_t : t = 1, 2, \ldots, n\}$ is *discrete white noise* (DWN) if the variables $w_1, w_2, \ldots, w_n$ are *independent* and *identically* distributed with a mean of zero. This implies that the variables all have the same variance σ^2 and $\mathrm{Cor}(w_i, w_j) = 0$ for all $i \neq j$. If, in addition, the variables also follow a normal distribution (i.e., $w_t \sim \mathrm{N}(0, \sigma^2)$) the series is called *Gaussian* white noise.

4.2.3 Simulation in R

A fitted time series model can be used to *simulate* data. Time series simulated using a model are sometimes called *synthetic* series to distinguish them from an observed historical series.

Simulation is useful for many reasons. For example, simulation can be used to generate plausible future scenarios and to construct confidence intervals for model parameters (sometimes called *bootstrapping*). In R, simulation is usually straightforward, and most standard statistical distributions are simulated using a function that has an abbreviated name for the distribution prefixed with an 'r' (for 'random').[1] For example, rnorm(100) is used to simulate 100 independent standard normal variables, which is equivalent to simulating a Gaussian white noise series of length 100 (Fig. 4.1).

```
> set.seed(1)
> w <- rnorm(100)
> plot(w, type = "l")
```

---

[1] Other prefixes are also available to calculate properties for standard distributions; e.g., the prefix 'd' is used to calculate the probability (density) function. See the R help (e.g., ?dnorm) for more details.

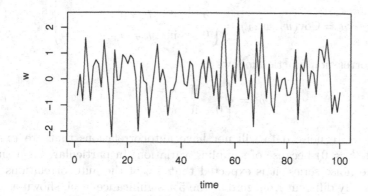

Fig. 4.1. Time plot of simulated Gaussian white noise series.

Simulation experiments in R can easily be repeated using the 'up' arrow on the keyboard. For this reason, it is sometimes preferable to put all the commands on one line, separated by ';', or to nest the functions; for example, a plot of a white noise series is given by `plot(rnorm(100), type="l")`.

The function `set.seed` is used to provide a starting point (or *seed*) in the simulations, thus ensuring that the simulations can be reproduced. If this function is left out, a different set of simulated data are obtained, although the underlying statistical properties remain unchanged. To see this, rerun the plot above a few times with and without `set.seed(1)`.

To illustrate by simulation how samples may differ from their underlying populations, consider the following histogram of a Gaussian white noise series. Type the following to view the plot (which is not shown in the text):

```
> x <- seq(-3,3, length = 1000)
> hist(rnorm(100), prob = T); points(x, dnorm(x), type = "l")
```

Repetitions of the last command, which can be obtained using the 'up' arrow on your keyboard, will show a range of different *sample* distributions that arise when the underlying distribution is normal. Distributions that depart from the plotted curve have arisen due to sampling variation.

4.2.4 Second-order properties and the correlogram

The second-order properties of a white noise series $\{w_t\}$ are an immediate consequence of the definition in §4.2.2. However, as they are needed so often in the derivation of the second-order properties for more complex models, we explicitly state them here:

$$\mu_w = 0$$

$$\gamma_k = \mathrm{Cov}(w_t, w_{t+k}) = \begin{cases} \sigma^2 & \text{if} \quad k = 0 \\ 0 & \text{if} \quad k \neq 0 \end{cases} \qquad (4.2)$$

The autocorrelation function follows as

$$\rho_k = \begin{cases} 1 & \text{if} \quad k = 0 \\ 0 & \text{if} \quad k \neq 0 \end{cases} \qquad (4.3)$$

Simulated white noise data will not have autocorrelations that are *exactly* zero (when $k \neq 0$) because of sampling variation. In particular, for a simulated white noise series, it is expected that 5% of the autocorrelations will be significantly different from zero at the 5% significance level, shown as dotted lines on the correlogram. Try repeating the following command to view a range of correlograms that could arise from an underlying white noise series. A typical plot, with one statistically significant autocorrelation, occurring at lag 7, is shown in Figure 4.2.

```
> set.seed(2)
> acf(rnorm(100))
```

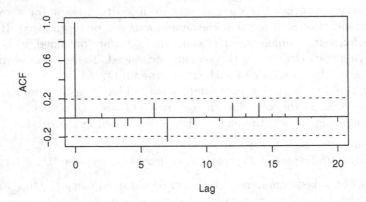

Fig. 4.2. Correlogram of a simulated white noise series. The underlying autocorrelations are all zero (except at lag 0); the statistically significant value at lag 7 is due to sampling variation.

4.2.5 Fitting a white noise model

A white noise series usually arises as a residual series after fitting an appropriate time series model. The correlogram generally provides sufficient evidence, provided the series is of a reasonable length, to support the conjecture that the residuals are well approximated by white noise.

The only parameter for a white noise series is the variance σ^2, which is estimated by the residual variance, adjusted by degrees of freedom, given in the computer output of the fitted model. If your analysis begins on data that are already approximately white noise, then only σ^2 needs to be estimated, which is readily achieved using the var function.

4.3 Random walks

4.3.1 Introduction

In Chapter 1, the exchange rate data were examined and found to exhibit stochastic trends. A random walk often provides a good fit to data with stochastic trends, although even better fits are usually obtained from more general model formulations, such as the ARIMA models of Chapter 7.

4.3.2 Definition

Let $\{x_t\}$ be a time series. Then $\{x_t\}$ is a random walk if

$$x_t = x_{t-1} + w_t \tag{4.4}$$

where $\{w_t\}$ is a white noise series. Substituting $x_{t-1} = x_{t-2} + w_{t-1}$ in Equation (4.4) and then substituting for x_{t-2}, followed by x_{t-3} and so on (a process known as 'back substitution') gives:

$$x_t = w_t + w_{t-1} + w_{t-2} + \dots \tag{4.5}$$

In practice, the series above will not be infinite but will start at some time $t = 1$. Hence,

$$x_t = w_1 + w_2 + \dots + w_t \tag{4.6}$$

Back substitution is used to define more complex time series models and also to derive second-order properties. The procedure occurs so frequently in the study of time series models that the following definition is needed.

4.3.3 The backward shift operator

The *backward shift* operator $\mathbf{B}$ is defined by

$$\mathbf{B}x_t = x_{t-1} \tag{4.7}$$

The backward shift operator is sometimes called the 'lag operator'. By repeatedly applying $\mathbf{B}$, it follows that

$$\mathbf{B}^n x_t = x_{t-n} \tag{4.8}$$

Using $\mathbf{B}$, Equation (4.4) can be rewritten as

$$x_t = \mathbf{B}x_t + w_t \Rightarrow (1 - \mathbf{B})x_t = w_t \Rightarrow x_t = (1 - \mathbf{B})^{-1}w_t$$

$$\Rightarrow x_t = (1 + \mathbf{B} + \mathbf{B}^2 + \ldots)w_t \Rightarrow x_t = w_t + w_{t-1} + w_{t-2} + \cdots$$

and Equation (4.5) is recovered.

4.3.4 Random walk: Second-order properties

The second-order properties of a random walk follow as

$$\left.\begin{array}{l} \mu_x = 0 \\[2mm] \gamma_k(t) = \mathrm{Cov}(x_t, x_{t+k}) = t\sigma^2 \end{array}\right\} \tag{4.9}$$

The covariance is a function of time, so the process is non-stationary. In particular, the variance is $t\sigma^2$ and so it increases without limit as t increases. It follows that a random walk is only suitable for short term predictions.

The time-varying autocorrelation function for $k > 0$ follows from Equation (4.9) as

$$\rho_k(t) = \frac{\mathrm{Cov}(x_t, x_{t+k})}{\sqrt{\mathrm{Var}(x_t)\mathrm{Var}(x_{t+k})}} = \frac{t\sigma^2}{\sqrt{t\sigma^2(t+k)\sigma^2}} = \frac{1}{\sqrt{1+k/t}} \tag{4.10}$$

so that, for large t with k considerably less than t, ρ_k is nearly 1. Hence, the correlogram for a random walk is characterised by positive autocorrelations that decay very slowly down from unity. This is demonstrated by simulation in §4.3.7.

4.3.5 Derivation of second-order properties*

Equation (4.6) is a finite sum of white noise terms, each with zero mean and variance σ^2. Hence, the mean of x_t is zero (Equation (4.9)). The autocovariance in Equation (4.9) can be derived using Equation (2.15) as follows:

$$\gamma_k(t) = \mathrm{Cov}(x_t, x_{t+k}) = \mathrm{Cov}\left(\sum_{i=1}^{t} w_i, \sum_{j=1}^{t+k} w_j\right) = \sum_{i=j}\mathrm{Cov}(w_i, w_j) = t\sigma^2$$

4.3.6 The difference operator

Differencing adjacent terms of a series can transform a non-stationary series to a stationary series. For example, if the series $\{x_t\}$ is a random walk, it is non-stationary. However, from Equation (4.4), the first-order differences of $\{x_t\}$ produce the stationary white noise series $\{w_t\}$ given by $x_t - x_{t-1} = w_t$.

Hence, differencing turns out to be a useful 'filtering' procedure in the study of non-stationary time series. The difference operator ∇ is defined by

$$\nabla x_t = x_t - x_{t-1} \qquad (4.11)$$

Note that $\nabla x_t = (1 - \mathbf{B})x_t$, so that ∇ can be expressed in terms of the backward shift operator $\mathbf{B}$. In general, higher-order differencing can be expressed as

$$\nabla^n = (1 - \mathbf{B})^n \qquad (4.12)$$

The proof of the last result is left to Exercise 7.

4.3.7 Simulation

It is often helpful to study a time series model by simulation. This enables the main features of the model to be observed in plots, so that when historical data exhibit similar features, the model may be selected as a potential candidate. The following commands can be used to simulate random walk data for x:

```
> x <- w <- rnorm(1000)
> for (t in 2:1000) x[t] <- x[t - 1] + w[t]
> plot(x, type = "l")
```

The first command above places a white noise series into w and uses this series to initialise x. The 'for' loop then generates the random walk using Equation (4.4) – the correspondence between the R code above and Equation (4.4) should be noted. The series is plotted and shown in Figure 4.3.[2]

A correlogram of the series is obtained from acf(x) and is shown in Figure 4.4 – a gradual decay in the correlations is evident in the figure, thus supporting the theoretical results in §4.3.4.

Throughout this book, we will often fit models to data that we have simulated and attempt to recover the underlying model parameters. At first sight, this might seem odd, given that the parameters are used to simulate the data so that we already know at the outset the values the parameters should take. However, the procedure is useful for a number of reasons. In particular, to be able to simulate data using a model requires that the model formulation be correctly understood. If the model is understood but incorrectly implemented, then the parameter estimates from the fitted model may deviate significantly from the underlying model values used in the simulation. Simulation can therefore help ensure that the model is both correctly understood and correctly implemented.

---

[2] To obtain the same simulation and plot, it is necessary to have run the previous code in §4.2.4 first, which sets the random number seed.

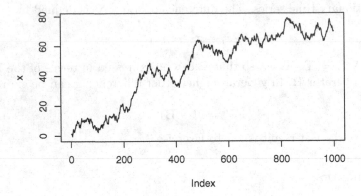

Fig. 4.3. Time plot of a simulated random walk. The series exhibits an increasing trend. However, this is purely stochastic and due to the high serial correlation.

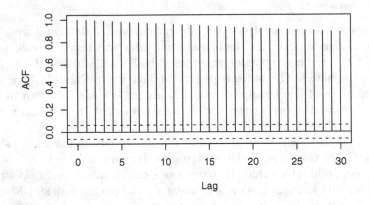

Fig. 4.4. The correlogram for the simulated random walk. A gradual decay from a high serial correlation is a notable feature of a random walk series.

4.4 Fitted models and diagnostic plots

4.4.1 Simulated random walk series

The first-order differences of a random walk are a white noise series, so the correlogram of the series of differences can be used to assess whether a given series is reasonably modelled as a random walk.

```
> acf(diff(x))
```

As can be seen in Figure 4.5, there are no obvious patterns in the correlogram, with only a couple of marginally statistically significant values. These significant values can be ignored because they are small in magnitude and about 5% of the values are expected to be statistically significant even when the underlying values are zero (§2.3). Thus, as expected, there is good evidence that the simulated series in x follows a random walk.

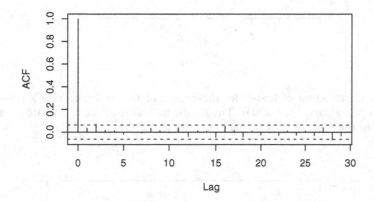

Fig. 4.5. Correlogram of differenced series. If a series follows a random walk, the differenced series will be white noise.

4.4.2 Exchange rate series

The correlogram of the first-order differences of the exchange rate data from §1.4.4 can be obtained from `acf(diff(Z.ts))` and is shown in Figure 4.6.

A significant value occurs at lag 1, suggesting that a more complex model may be needed, although the lack of any other significant values in the correlogram does suggest that the random walk provides a good approximation for the series (Fig. 4.6). An additional term can be added to the random walk model using the Holt-Winters procedure, allowing the parameter β to be non-zero but still forcing the seasonal term γ to be zero:

```
> Z.hw <- HoltWinters(Z.ts, alpha = 1, gamma = 0)
> acf(resid(Z.hw))
```

Figure 4.7 shows the correlogram of the residuals from the fitted Holt-Winters model. This correlogram is more consistent with a hypothesis that the residual series is white noise (Fig. 4.7). Using Equation (3.21), with the parameter estimates obtained from `Z.hw$alpha` and `Z.hw$beta`, the fitted model can be expressed as

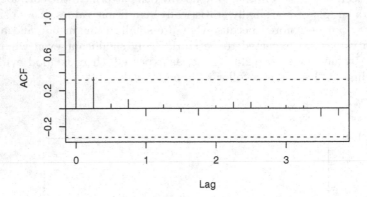

Fig. 4.6. Correlogram of first-order differences of the exchange rate series (UK pounds to NZ dollars, 1991–2000). The significant value at lag 1 indicates that an extension of the random walk model is needed for this series.

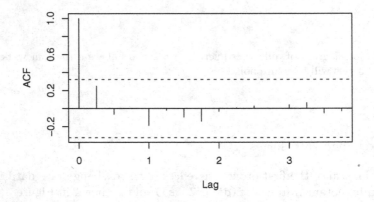

Fig. 4.7. The correlogram of the residuals from the fitted Holt-Winters model for the exchange rate series (UK pounds to NZ dollars, 1991–2000). There are no significant correlations in the residual series, so the model provides a reasonable approximation to the exchange rate data.

$$\left.\begin{array}{l} x_t = x_{t-1} + b_{t-1} + w_t \\ b_{t-1} = 0.167(x_{t-1} - x_{t-2}) + 0.833b_{t-2} \end{array}\right\} \qquad (4.13)$$

where $\{w_t\}$ is white noise with zero mean.

After some algebra, Equations (4.13) can be expressed as one equation in terms of the backward shift operator:

$$(1 - 0.167\mathbf{B} + 0.167\mathbf{B}^2)(1 - \mathbf{B})x_t = w_t \qquad (4.14)$$

Equation (4.14) is a special case – the *integrated autoregressive* model – within the important class of models known as ARIMA models (Chapter 7). The proof of Equation (4.14) is left to Exercise 8.

4.4.3 Random walk with drift

Company stockholders generally expect their investment to increase in value despite the volatility of financial markets. The random walk model can be adapted to allow for this by including a *drift* parameter δ.

$$x_t = x_{t-1} + \delta + w_t$$

Closing prices (US dollars) for Hewlett-Packard Company stock for 672 trading days up to June 7, 2007 are read into R and plotted (see the code below and Fig. 4.8). The lag 1 differences are calculated using diff() and plotted in Figure 4.9. The correlogram of the differences is in Figure 4.10, and they appear to be well modelled as white noise. The mean of the differences is 0.0399, and this is our estimate of the drift parameter. The standard deviation of the 671 differences is 0.460, and an approximate 95% confidence interval for the drift parameter is [0.004, 0.075]. Since this interval does not include 0, we have evidence of a positive drift over this period.

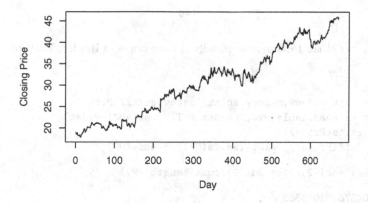

Fig. 4.8. Daily closing prices of Hewlett-Packard stock.

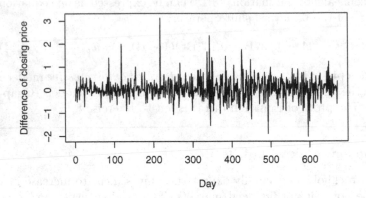

Fig. 4.9. Lag 1 differences of daily closing prices of Hewlett-Packard stock.

Series DP

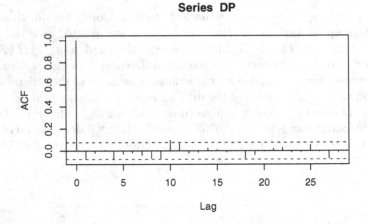

Fig. 4.10. Acf of lag 1 differences of daily closing prices of Hewlett-Packard stock.

```
> www <- "http://www.massey.ac.nz/~pscowper/ts/HP.txt"
> HP.dat <- read.table(www, header = T) ; attach(HP.dat)
> plot (as.ts(Price))
> DP <- diff(Price) ; plot (as.ts(DP)) ; acf (DP)

> mean(DP) + c(-2, 2) * sd(DP)/sqrt(length(DP))

[1] 0.004378 0.075353
```

4.5 Autoregressive models

4.5.1 Definition

The series $\{x_t\}$ is an autoregressive process of order p, abbreviated to AR(p), if

$$x_t = \alpha_1 x_{t-1} + \alpha_2 x_{t-2} + \ldots + \alpha_p x_{t-p} + w_t \qquad (4.15)$$

where $\{w_t\}$ is white noise and the α_i are the model parameters with $\alpha_p \neq 0$ for an order p process. Equation (4.15) can be expressed as a polynomial of order p in terms of the backward shift operator:

$$\theta_p(\mathbf{B})x_t = (1 - \alpha_1 \mathbf{B} - \alpha_2 \mathbf{B}^2 - \ldots - \alpha_p \mathbf{B}^p)x_t = w_t \qquad (4.16)$$

The following points should be noted:

(a) The random walk is the special case AR(1) with $\alpha_1 = 1$ (see Equation (4.4)).
(b) The exponential smoothing model is the special case $\alpha_i = \alpha(1 - \alpha)^i$ for $i = 1, 2, \ldots$ and $p \to \infty$.
(c) The model is a regression of x_t on past terms from the same series; hence the use of the term 'autoregressive'.
(d) A prediction at time t is given by

$$\hat{x}_t = \alpha_1 x_{t-1} + \alpha_2 x_{t-2} + \ldots + \alpha_p x_{t-p} \qquad (4.17)$$

(e) The model parameters can be estimated by minimising the sum of squared errors.

4.5.2 Stationary and non-stationary AR processes

The equation $\theta_p(\mathbf{B}) = 0$, where $\mathbf{B}$ is formally treated as a number (real or complex), is called the characteristic equation. The roots of the characteristic equation (i.e., the polynomial $\theta_p(\mathbf{B})$ from Equation (4.16)) must *all* exceed unity in *absolute* value for the process to be *stationary*. Notice that the random walk has $\theta = 1 - \mathbf{B}$ with root $\mathbf{B} = 1$ and is *non-stationary*. The following four examples illustrate the procedure for determining whether an AR process is stationary or non-stationary:

1. The AR(1) model $x_t = \frac{1}{2}x_{t-1} + w_t$ is stationary because the root of $1 - \frac{1}{2}\mathbf{B} = 0$ is $\mathbf{B} = 2$, which is greater than 1.
2. The AR(2) model $x_t = x_{t-1} - \frac{1}{4}x_{t-2} + w_t$ is stationary. The proof of this result is obtained by first expressing the model in terms of the backward shift operator $\frac{1}{4}(\mathbf{B}^2 - 4\mathbf{B} + 4)x_t = w_t$; i.e., $\frac{1}{4}(\mathbf{B} - 2)^2 x_t = w_t$. The roots of the polynomial are given by solving $\theta(\mathbf{B}) = \frac{1}{4}(\mathbf{B} - 2)^2 = 0$ and are therefore obtained as $\mathbf{B} = 2$. As the roots are greater than unity this AR(2) model is stationary.

3. The model $x_t = \frac{1}{2}x_{t-1} + \frac{1}{2}x_{t-2} + w_t$ is non-stationary because one of the roots is unity. To prove this, first express the model in terms of the backward shift operator $-\frac{1}{2}(\mathbf{B}^2 + \mathbf{B} - 2)x_t = w_t$; i.e., $-\frac{1}{2}(\mathbf{B} - 1)(\mathbf{B} + 2)x_t = w_t$. The polynomial $\theta(\mathbf{B}) = -\frac{1}{2}(\mathbf{B} - 1)(\mathbf{B} + 2)$ has roots $\mathbf{B} = 1, -2$. As there is a *unit root* ($\mathbf{B} = 1$), the model is *non-stationary*. Note that the other root ($\mathbf{B} = -2$) exceeds unity in *absolute* value, so only the presence of the unit root makes this process non-stationary.

4. The AR(2) model $x_t = -\frac{1}{4}x_{t-2} + w_t$ is stationary because the roots of $1 + \frac{1}{4}\mathbf{B}^2 = 0$ are $\mathbf{B} = \pm 2i$, which are complex numbers with $i = \sqrt{-1}$, each having an absolute value of 2 exceeding unity.

The R function `polyroot` finds zeros of polynomials and can be used to find the roots of the characteristic equation to check for stationarity.

4.5.3 Second-order properties of an AR(1) model

From Equation (4.15), the AR(1) process is given by

$$x_t = \alpha x_{t-1} + w_t \tag{4.18}$$

where $\{w_t\}$ is a white noise series with mean zero and variance σ^2. It can be shown (§4.5.4) that the second-order properties follow as

$$\left. \begin{array}{l} \mu_x = 0 \\ \gamma_k = \alpha^k \sigma^2 / (1 - \alpha^2) \end{array} \right\} \tag{4.19}$$

4.5.4 Derivation of second-order properties for an AR(1) process*

Using $\mathbf{B}$, a stable AR(1) process ($|\alpha| < 1$) can be written as

$$\left. \begin{array}{l} (1 - \alpha \mathbf{B})x_t = w_t \\ \Rightarrow x_t = (1 - \alpha \mathbf{B})^{-1} w_t \\ \quad = w_t + \alpha w_{t-1} + \alpha^2 w_{t-2} + \ldots = \sum_{i=0}^{\infty} \alpha^i w_{t-i} \end{array} \right\} \tag{4.20}$$

Hence, the mean is given by

$$E(x_t) = E\left(\sum_{i=0}^{\infty} \alpha^i w_{t-i} \right) = \sum_{i=0}^{\infty} \alpha^i E(w_{t-i}) = 0$$

and the autocovariance follows as

$$\begin{aligned} \gamma_k = \mathrm{Cov}(x_t, x_{t+k}) &= \mathrm{Cov}\left(\sum_{i=0}^{\infty} \alpha^i w_{t-i}, \sum_{j=0}^{\infty} \alpha^j w_{t+k-j} \right) \\ &= \sum_{j=k+i} \alpha^i \alpha^j \mathrm{Cov}(w_{t-i}, w_{t+k-j}) \\ &= \alpha^k \sigma^2 \sum_{i=0}^{\infty} \alpha^{2i} = \alpha^k \sigma^2 / (1 - \alpha^2) \end{aligned}$$

using Equations (2.15) and (4.2).

4.5.5 Correlogram of an AR(1) process

From Equation (4.19), the autocorrelation function follows as

$$\rho_k = \alpha^k \qquad (k \geq 0) \qquad\qquad (4.21)$$

where $|\alpha| < 1$. Thus, the correlogram decays to zero more rapidly for small α. The following example gives two correlograms for positive and negative values of α, respectively (Fig. 4.11):

```
> rho <- function(k, alpha) alpha^k
> layout(1:2)
> plot(0:10, rho(0:10, 0.7), type = "b")
> plot(0:10, rho(0:10, -0.7), type = "b")
```

Try experimenting using other values for α. For example, use a small value of α to observe a more rapid decay to zero in the correlogram.

4.5.6 Partial autocorrelation

From Equation (4.21), the autocorrelations are non-zero for all lags even though in the underlying model x_t only depends on the previous value x_{t-1} (Equation (4.18)). The *partial autocorrelation* at lag k is the correlation that results after removing the effect of any correlations due to the terms at shorter lags. For example, the partial autocorrelation of an AR(1) process will be zero for all lags greater than 1. In general, the partial autocorrelation at lag k is the kth coefficient of a fitted AR(k) model; if the underlying process is AR(p), then the coefficients α_k will be zero for all $k > p$. Thus, an AR(p) process has a correlogram of partial autocorrelations that is zero after lag p. Hence, a plot of the estimated partial autocorrelations can be useful when determining the order of a suitable AR process for a time series. In R, the function `pacf` can be used to calculate the partial autocorrelations of a time series and produce a plot of the partial autocorrelations against lag (the 'partial correlogram').

4.5.7 Simulation

An AR(1) process can be simulated in R as follows:

```
> set.seed(1)
> x <- w <- rnorm(100)
> for (t in 2:100) x[t] <- 0.7 * x[t - 1] + w[t]
> plot(x, type = "l")
> acf(x)
> pacf(x)
```

The resulting plots of the simulated data are shown in Figure 4.12 and give one possible realisation of the model. The partial correlogram has no significant correlations except the value at lag 1, as expected (Fig. 4.12c – note that the

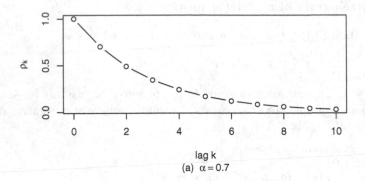

(a) $\alpha = 0.7$

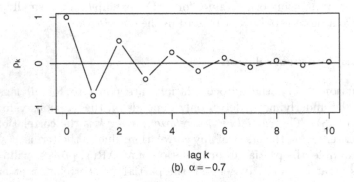

(b) $\alpha = -0.7$

Fig. 4.11. Example correlograms for two autoregressive models: (a) $x_t = 0.7x_{t-1} + w_t$; (b) $x_t = -0.7x_{t-1} + w_t$.

pacf starts at lag 1, whilst the acf starts at lag 0). The difference between the correlogram of the underlying model (Fig. 4.11a) and the sample correlogram of the simulated series (Fig. 4.12b) shows discrepancies that have arisen due to sampling variation. Try repeating the commands above several times to obtain a range of possible sample correlograms for an AR(1) process with underlying parameter $\alpha = 0.7$. You are asked to investigate an AR(2) process in Exercise 4.

4.6 Fitted models

4.6.1 Model fitted to simulated series

An AR(p) model can be fitted to data in R using the **ar** function. In the code below, the autoregressive model **x.ar** is fitted to the simulated series of the last section and an approximate 95% confidence interval for the underlying

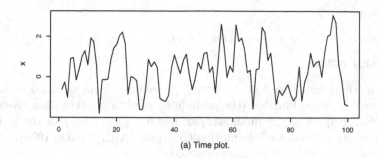

(a) Time plot.

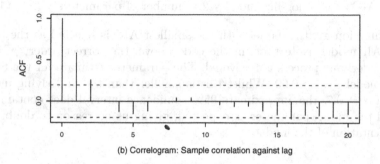

(b) Correlogram: Sample correlation against lag

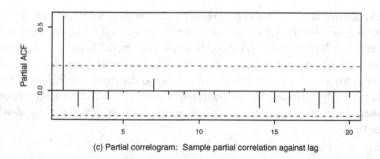

(c) Partial correlogram: Sample partial correlation against lag

Fig. 4.12. A simulated AR(1) process, $x_t = 0.7x_{t-1} + w_t$. Note that in the partial correlogram (c) only the first lag is significant, which is usually the case when the underlying process is AR(1).

parameter is given, where the (asymptotic) variance of the parameter estimate is extracted using x.ar$asy.var:

```
> x.ar <- ar(x, method = "mle")
> x.ar$order

[1] 1

> x.ar$ar
```

```
[1] 0.601
```

```
> x.ar$ar + c(-2, 2) * sqrt(x.ar$asy.var)
```

```
[1] 0.4404 0.7615
```

The method "mle" used in the fitting procedure above is based on max-imising the likelihood function (the probability of obtaining the data given the model) with respect to the unknown parameters. The order p of the process is chosen using the Akaike Information Criterion (AIC; Akaike, 1974), which penalises models with too many parameters:

$$\text{AIC} = -2 \times \text{log-likelihood} + 2 \times \text{number of parameters} \qquad (4.22)$$

In the function ar, the model with the smallest AIC is selected as the best-fitting AR model. Note that, in the code above, the correct order ($p = 1$) of the underlying process is recovered. The parameter estimate for the fitted AR(1) model is $\hat{\alpha} = 0.60$. Whilst this is smaller than the underlying model value of $\alpha = 0.7$, the approximate 95% confidence interval does contain the value of the model parameter as expected, giving us no reason to doubt the implementation of the model.

4.6.2 Exchange rate series: Fitted AR model

An AR(1) model is fitted to the exchange rate series, and the upper bound of the confidence interval for the parameter includes 1. This indicates that there would not be sufficient evidence to reject the hypothesis $\alpha = 1$, which is consistent with the earlier conclusion that a random walk provides a good ap-proximation for this series. However, simulated data from models with values of $\alpha > 1$, formally included in the confidence interval below, exhibit exponen-tially unstable behaviour and are not credible models for the New Zealand exchange rate.

```
> Z.ar <- ar(Z.ts)
> mean(Z.ts)
```

```
[1] 2.823
```

```
> Z.ar$order
```

```
[1] 1
```

```
> Z.ar$ar
```

```
[1] 0.8903
```

```
> Z.ar$ar + c(-2, 2) * sqrt(Z.ar$asy.var)
```

```
[1] 0.7405 1.0400
```

```
> acf(Z.ar$res[-1])
```

In the code above, a "−1" is used in the vector of residuals to remove the first item from the residual series (Fig. 4.13). (For a fitted AR(1) model, the first item has no predicted value because there is no observation at $t = 0$; in general, the first p values will be 'not available' (NA) in the residual series of a fitted AR(p) model.)

By default, the mean is subtracted before the parameters are estimated, so a predicted value $\hat{z}_t$ at time t based on the output above is given by

$$\hat{z}_t = 2.8 + 0.89(z_{t-1} - 2.8) \tag{4.23}$$

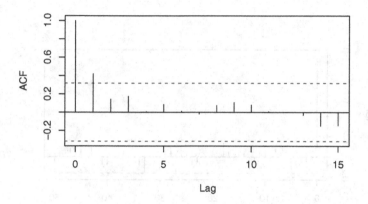

Fig. 4.13. The correlogram of residual series for the AR(1) model fitted to the exchange rate data.

4.6.3 Global temperature series: Fitted AR model

The global temperature series was introduced in §1.4.5, where it was apparent that the data exhibited an increasing trend after 1970, which may be due to the 'greenhouse effect'. Sceptics may claim that the apparent increasing trend can be dismissed as a transient stochastic phenomenon. For their claim to be consistent with the time series data, it should be possible to model the trend without the use of deterministic functions.

Consider the following AR model fitted to the mean annual temperature series:

```
> www = "http://www.massey.ac.nz/~pscowper/ts/global.dat"
> Global = scan(www)
> Global.ts = ts(Global, st = c(1856, 1), end = c(2005, 12),
      fr = 12)
```

```
> Global.ar <- ar(aggregate(Global.ts, FUN = mean), method = "mle")
> mean(aggregate(Global.ts, FUN = mean))

[1] -0.1383

> Global.ar$order

[1] 4

> Global.ar$ar

[1] 0.58762 0.01260 0.11117 0.26764

> acf(Global.ar$res[-(1:Global.ar$order)], lag = 50)
```

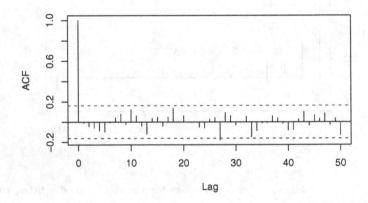

Fig. 4.14. The correlogram of the residual series for the AR(4) model fitted to the annual global temperature series. The correlogram is approximately white noise so that, in the absence of further information, a simple stochastic model can 'explain' the correlation and trends in the series.

Based on the output above a predicted mean annual temperature $\hat{x}_t$ at time t is given by

$$
\begin{aligned}
\hat{x}_t = -0.14 + 0.59(x_{t-1} + 0.14) + 0.013(x_{t-2} + 0.14) \\
+0.11(x_{t-3} + 0.14) + 0.27(x_{t-4} + 0.14)
\end{aligned}
\tag{4.24}
$$

The correlogram of the residuals has only one (marginally) significant value at lag 27, so the underlying residual series could be white noise (Fig. 4.14). Thus the fitted AR(4) model (Equation (4.24)) provides a good fit to the data. As the AR model has no deterministic trend component, the trends in the data can be explained by serial correlation and random variation, implying that it is possible that these trends are stochastic (or could arise from a purely stochastic process). Again we emphasise that this does not imply that there is no underlying reason for the trends. If a valid scientific explanation is known, such as a link with the increased use of fossil fuels, then this information would clearly need to be included in any future forecasts of the series.

4.7 Summary of R commands

set.seed	sets a seed for the random number generator enabling a simulation to be reproduced
rnorm	simulates Gaussian white noise series
diff	creates a series of first-order differences
ar	gets the best fitting AR(p) model
pacf	extracts partial autocorrelations and partial correlogram
polyroot	extracts the roots of a polynomial
resid	extracts the residuals from a fitted model

4.8 Exercises

1. Simulate discrete white noise from an exponential distribution and plot the histogram and the correlogram. For example, you can use the R command w <- rexp(1000)-1 for exponential white noise. Comment on the plots.

2. a) Simulate time series of length 100 from an AR(1) model with α equal to $-0.9, -0.5, 0.5$, and 0.9. Estimate the parameter of each model and make predictions for 1 to 10 steps ahead.
   b) Simulate time series of length 100 from an AR(1) model with α equal to $1.01, 1.02$, and 1.05. Estimate the parameters of these models.

3. An AR(1) model with a non-zero mean μ can be expressed by either $x_t - \mu = \alpha(x_{t-1} - \mu) + w_t$ or $x_t = \alpha_0 + \alpha_1 x_{t-1} + w_t$.
   a) What is the relationship between the parameters μ and α and the parameters α_0 and α_1?
   b) Deduce a similar relationship for an AR(2) process with mean μ.

4. a) Simulate a time series of length 1000 for the following model, giving appropriate R code and placing the simulated data in a vector x:

$$x_t = \frac{5}{6}x_{t-1} - \frac{1}{6}x_{t-2} + w_t \qquad (4.25)$$

   b) Plot the correlogram and partial correlogram for the simulated data. Comment on the plots.

c) Fit an AR model to the data in x giving the parameter estimates and order of the fitted AR process.

d) Construct 95% confidence intervals for the parameter estimates of the fitted model. Do the model parameters fall within the confidence intervals? Explain your results.

e) Is the model in Equation (4.25) stationary or non-stationary? Justify your answer.

f) Plot the correlogram of the residuals of the fitted model, and comment on the plot.

5. a) Show that the series $\{x_t\}$ given by $x_t = \frac{3}{2}x_{t-1} - \frac{1}{2}x_{t-2} + w_t$ is non-stationary.

b) Write down the model for $\{y_t\}$, where $y_t = \nabla x_t$. Show that $\{y_t\}$ is stationary.

c) Simulate a series of 1000 values for $\{x_t\}$, placing the simulated data in x, and use these simulated values to produce a series of 999 values for $\{y_t\}$, placing this series in the vector y.

d) Fit an AR model to y. Give the fitted model parameter estimates and a 95% confidence interval for the underlying model parameters based on these estimates. Compare the confidence intervals to the parameters used to simulate the data and explain the results.

e) Plot the correlogram of the residuals of the fitted model and comment.

6. a) Refit the AR(4) model of §4.6.3 to the annual mean global temperature series, and using the fitted model create a series of predicted values from $t = 2$ to the last value in the series (using Equation (4.24)). Create a residual series from the difference between the predicted value and the observed value, and verify that within machine accuracy your residual series is identical to the series extracted from the fitted model in R.

b) Plot a correlogram and partial correlogram for the mean annual temperature series. Comment on the plots.

c) Use the predict function in R to forecast 100 years of future values for the annual global temperature series using the fitted AR(4) model (Equation (4.24)) of §4.6.3.

d) Create a time plot of the mean annual temperature series and add the 100-year forecasts to the plot (use a different colour or symbol for the forecasts).

e) Add a line representing the overall mean global temperature. Comment on the final plot and any potential inadequacies in the fitted model.

7. Prove Equation (4.12) by mathematical induction as follows. (i) First, show that if Equation (4.12) holds for $n = k$, then it also holds for $n =$

$k + 1$. (ii) Next, show that Equation (4.12) holds for the case $n = 2$ and hence (from i) holds for all n.

8. Prove Equation (4.14). [Hint: Express the two equations in (4.13) in terms of the backward shift operator and then substitute for b_n.]

5

Regression

5.1 Purpose

Trends in time series can be classified as *stochastic* or *deterministic*. We may consider a trend to be stochastic when it shows inexplicable changes in direction, and we attribute apparent transient trends to high serial correlation with random error. Trends of this type, which are common in financial series, can be simulated in R using models such as the random walk or autoregressive process (Chapter 4). In contrast, when we have some plausible physical explanation for a trend we will usually wish to model it in some deterministic manner. For example, a deterministic increasing trend in the data may be related to an increasing population, or a regular cycle may be related to a known seasonal frequency. Deterministic trends and seasonal variation can be modelled using regression.

The practical difference between stochastic and deterministic trends is that we extrapolate the latter when we make forecasts. We justify short-term extrapolation by claiming that underlying trends will usually change slowly in comparison with the forecast lead time. For the same reason, short-term extrapolation should be based on a line, maybe fitted to the more recent data only, rather than a high-order polynomial.

In this chapter various regression models are studied that are suitable for a time series analysis of data that contain deterministic trends and regular seasonal changes. We begin by looking at linear models for trends and then introduce regression models that account for seasonal variation using indicator and harmonic variables. Regression models can also include explanatory variables. The logarithmic transformation, which is often used to stabilise the variance, is also considered.

Time series regression usually differs from a standard regression analysis because the residuals form a time series and therefore tend to be serially correlated. When this correlation is positive, the estimated standard errors of the parameter estimates, read from the computer output of a standard regression analysis, will tend to be less than their true value. This will lead

P.S.P. Cowpertwait and A.V. Metcalfe, *Introductory Time Series with R*,
Use R, DOI 10.1007/978-0-387-88698-5_5,
© Springer Science+Business Media, LLC 2009

to erroneously high statistical significance being attributed to statistical tests in standard computer output (the p values will be smaller than they should be). Presenting correct statistical evidence is important. For example, an environmental protection group could be undermined by allegations that it is falsely claiming statistically significant trends. In this chapter, generalised least squares is used to obtain improved estimates of the standard error to account for autocorrelation in the residual series.

5.2 Linear models

5.2.1 Definition

A model for a time series $\{x_t : t = 1, \ldots n\}$ is *linear* if it can be expressed as

$$x_t = \alpha_0 + \alpha_1 u_{1,t} + \alpha_2 u_{2,t} + \ldots + \alpha_m u_{m,t} + z_t \tag{5.1}$$

where $u_{i,t}$ is the value of the ith predictor (or explanatory) variable at time t ($i = 1, \ldots, m; t = 1, \ldots, n$), z_t is the error at time t, and $\alpha_0, \alpha_1, \ldots, \alpha_m$ are model parameters, which can be estimated by least squares. Note that the errors form a time series $\{z_t\}$, with mean 0, that does not have to be Gaussian or white noise. An example of a linear model is the pth-order polynomial function of t:

$$x_t = \alpha_0 + \alpha_1 t + \alpha_2 t^2 \ldots + \alpha_p t^p + z_t \tag{5.2}$$

The predictor variables can be written $u_{i,t} = t^i$ ($i = 1, \ldots, p$). The term 'linear' is a reference to the summation of model parameters, each multiplied by a single predictor variable.

A simple special case of a linear model is the straight-line model obtained by putting $p = 1$ in Equation (5.2): $x_t = \alpha_0 + \alpha_1 t + z_t$. In this case, the value of the line at time t is the trend m_t. For the more general polynomial, the trend at time t is the value of the underlying polynomial evaluated at t, so in Equation (5.2) the trend is $m_t = \alpha_0 + \alpha_1 t + \alpha_2 t^2 \ldots + \alpha_p t^p$.

Many non-linear models can be transformed to linear models. For example, the model $x_t = e^{\alpha_0 + \alpha_1 t + z_t}$ for the series $\{x_t\}$ can be transformed by taking natural logarithms to obtain a linear model for the series $\{y_t\}$:

$$y_t = \log x_t = \alpha_0 + \alpha_1 t + z_t \tag{5.3}$$

In Equation (5.3), standard least squares regression could then be used to fit a linear model (i.e., estimate the parameters α_0 and α_1) and make predictions for y_t. To make predictions for x_t, the inverse transform needs to be applied to y_t, which in this example is $\exp(y_t)$. However, this usually has the effect of biasing the forecasts of mean values, and we discuss correction factors in §5.10.

Natural processes that generate time series are not expected to be precisely linear, but linear approximations are often adequate. However, we are not restricted to linear models, and the Bass model (§3.3) is an example of a non-linear model, which we fitted using the non-linear least squares function nls.

5.2.2 Stationarity

Linear models for time series are non-stationary when they include functions of time. Differencing can often transform a non-stationary series with a deterministic trend to a stationary series. For example, if the time series $\{x_t\}$ is given by the straight-line function plus white noise $x_t = \alpha_0 + \alpha_1 t + z_t$, then the first-order differences are given by

$$\nabla x_t = x_t - x_{t-1} = z_t - z_{t-1} + \alpha_1 \qquad (5.4)$$

Assuming the error series $\{z_t\}$ is stationary, the series $\{\nabla x_t\}$ is stationary as it is not a function of t. In §4.3.6 we found that first-order differencing can transform a non-stationary series with a stochastic trend (the random walk) to a stationary series. Thus, differencing can remove both stochastic and deterministic trends from time series. If the underlying trend is a polynomial of order m, then mth-order differencing is required to remove the trend.

   Notice that differencing the straight-line function plus white noise leads to a different stationary time series than subtracting the trend. The latter gives white noise, whereas differencing gives a series of consecutive white noise terms (which is an example of an MA process, described in Chapter 6).

5.2.3 Simulation

In time series regression, it is common for the error series $\{z_t\}$ in Equation (5.1) to be autocorrelated. In the code below a time series with an increasing straight-line trend $(50 + 3t)$ with autocorrelated errors is simulated and plotted:

```
> set.seed(1)
> z <- w <- rnorm(100, sd = 20)
> for (t in 2:100) z[t] <- 0.8 * z[t - 1] + w[t]
> Time <- 1:100
> x <- 50 + 3 * Time + z
> plot(x, xlab = "time", type = "l")
```

The model for the code above can be expressed as $x_t = 50 + 3t + z_t$, where $\{z_t\}$ is the AR(1) process $z_t = 0.8z_{t-1} + w_t$ and $\{w_t\}$ is Gaussian white noise with $\sigma = 20$. A time plot of a realisation of $\{x_t\}$ is given in Figure 5.1.

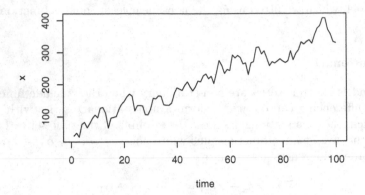

Fig. 5.1. Time plot of a simulated time series with a straight-line trend and AR(1) residual errors.

5.3 Fitted models

5.3.1 Model fitted to simulated data

Linear models are usually fitted by minimising the sum of squared errors, $\sum z_t^2 = \sum (x_t - \alpha_0 - \alpha_1 u_{1,t} - \ldots - \alpha_m u_{m,t})^2$, which is achieved in R using the function lm:

```
> x.lm <- lm(x ~ Time)
> coef(x.lm)

(Intercept)        Time
      58.55        3.06

> sqrt(diag(vcov(x.lm)))

(Intercept)        Time
     4.8801      0.0839
```

In the code above, the estimated parameters of the linear model are extracted using coef. Note that, as expected, the estimates are close to the underlying parameter values of 50 for the intercept and 3 for the slope. The standard errors are extracted using the square root of the diagonal elements obtained from vcov, although these standard errors are likely to be underestimated because of autocorrelation in the residuals. The function summary can also be used to obtain this information but tends to give additional information, for example t-tests, which may be incorrect for a time series regression analysis due to autocorrelation in the residuals.

After fitting a regression model, we should consider various diagnostic plots. In the case of time series regression, an important diagnostic plot is the correlogram of the residuals:

```
> acf(resid(x.lm))
> pacf(resid(x.lm))
```

As expected, the residual time series is autocorrelated (Fig. 5.2). In Figure 5.3, only the lag 1 partial autocorrelation is significant, which suggests that the residual series follows an AR(1) process. Again this should be as expected, given that an AR(1) process was used to simulate these residuals.

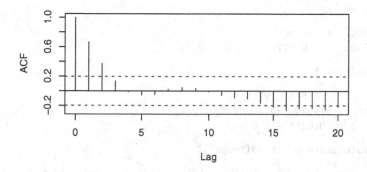

Fig. 5.2. Residual correlogram for the fitted straight-line model.

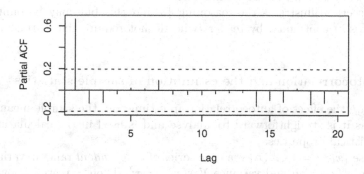

Fig. 5.3. Residual partial correlogram for the fitted straight-line model.

5.3.2 Model fitted to the temperature series (1970–2005)

In §1.4.5, we extracted temperatures for the period 1970–2005. The following regression model is fitted to the global temperature over this period, and approximate 95% confidence intervals are given for the parameters using `confint`. The explanatory variable is the time, so the function `time` is used to extract the 'times' from the `ts` temperature object.

```
> www <- "http://www.massey.ac.nz/~pscowper/ts/global.dat"
> Global <- scan(www)
> Global.ts <- ts(Global, st = c(1856, 1), end = c(2005,
    12), fr = 12)
> temp <- window(Global.ts, start = 1970)
> temp.lm <- lm(temp ~ time(temp))
> coef(temp.lm)

(Intercept)  time(temp)
  -34.9204      0.0177

> confint(temp.lm)

                2.5 %    97.5 %
(Intercept) -37.2100  -32.6308
time(temp)    0.0165    0.0188

> acf(resid(lm(temp ~ time(temp))))
```

The confidence interval for the slope does not contain zero, which would provide statistical evidence of an increasing trend in global temperatures if the autocorrelation in the residuals is negligible. However, the residual series is positively autocorrelated at shorter lags (Fig. 5.4), leading to an underestimate of the standard error and too narrow a confidence interval for the slope.

Intuitively, the positive correlation between consecutive values reduces the effective record length because similar values will tend to occur together. The following section illustrates the reasoning behind this but may be omitted, without loss of continuity, by readers who do not require the mathematical details.

5.3.3 Autocorrelation and the estimation of sample statistics*

To illustrate the effect of autocorrelation in estimation, the sample mean will be used, as it is straightforward to analyse and is used in the calculation of other statistical properties.

Suppose $\{x_t : t = 1, \ldots, n\}$ is a time series of *independent* random variables with mean $E(x_t) = \mu$ and variance $\text{Var}(x_t) = \sigma^2$. Then it is well known in the study of random samples that the sample mean $\bar{x} = \sum_{t=1}^{n} x_t/n$ has mean $E(\bar{x}) = \mu$ and variance $\text{Var}(\bar{x}) = \sigma^2/n$ (or standard error $\sigma/\sqrt{n}$). Now let $\{x_t : t = 1, \ldots, n\}$ be a stationary time series with $E(x_t) = \mu$, $\text{Var}(x_t) = \sigma^2$, and autocorrelation function $\text{Cor}(x_t, x_{t+k}) = \rho_k$. Then the variance of the sample mean is given by

$$\text{Var}(\bar{x}) = \frac{\sigma^2}{n}\left[1 + 2\sum_{k=1}^{n-1}(1 - k/n)\rho_k\right] \tag{5.5}$$

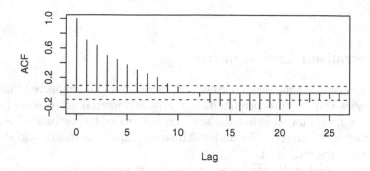

Fig. 5.4. Residual correlogram for the regression model fitted to the global temperature series (1970–2005).

In Equation (5.5) the variance σ^2/n for an independent random sample arises as the special case where $\rho_k = 0$ for all $k > 0$. If $\rho_k > 0$, then $\mathrm{Var}(\bar{x}) > \sigma^2/n$ and the resulting estimate of μ is less accurate than that obtained from a random (independent) sample of the same size. Conversely, if $\rho_k < 0$, then the variance of the estimate may actually be smaller than the variance obtained from a random sample of the same size. This latter result is due to the tendency for a value above the mean to be followed by a value below the mean, thus providing a more efficient estimate of the overall mean level. Conversely, for a positive correlation, values are more likely to persist above or below the mean, resulting in a less efficient estimate of the overall mean. Thus, for a positively correlated series, a larger sample would be needed to achieve the same level of accuracy in the estimate of μ obtained from a sample of negatively (or zero) correlated series. Equation (5.5) can be proved using Equation (2.15) and the properties of variance:

$$
\begin{aligned}
\mathrm{Var}\left(\bar{x}\right) &= \mathrm{Var}\left[(x_1 + x_2 + \cdots + x_n)/n\right] = \mathrm{Var}\left(x_1 + x_2 + \cdots + x_n\right)/n^2 \\
&= n^{-2}\mathrm{Cov}\left(\textstyle\sum_{i=1}^{n} x_i, \sum_{j=1}^{n} x_j\right) = n^{-2}\sum_{i=1}^{n}\sum_{j=1}^{n}\mathrm{Cov}\left(x_i, x_j\right) \\
&= n^{-2}\big[\gamma_0 \quad + \gamma_1 \quad + \cdots + \gamma_{n-2} + \gamma_{n-1} + \\
&\qquad\quad \gamma_1 \quad + \gamma_0 \quad + \cdots + \gamma_{n-3} + \gamma_{n-2} + \\
&\qquad\qquad\qquad \vdots \qquad\qquad\qquad\qquad \vdots \\
&\qquad \gamma_{n-2} + \gamma_{n-3} + \cdots + \gamma_2 \quad + \gamma_1 \quad + \\
&\qquad \gamma_{n-1} + \gamma_{n-2} + \cdots + \gamma_1 \quad + \gamma_0\big] \\
&= n^{-2}\big[n\gamma_0 + 2\textstyle\sum_{k=1}^{n-1}(n-k)\gamma_k\big]
\end{aligned}
$$

Equation (5.5) follows after substituting $\gamma_0 = \sigma^2$ and $\rho_k = \gamma_k/\sigma^2$ in the last line above.

5.4 Generalised least squares

We have seen that in time series regression it is common and expected that the residual series will be autocorrelated. For a positive serial correlation in the residual series, this implies that the standard errors of the estimated regression parameters are likely to be underestimated (Equation (5.5)), and should therefore be corrected.

A fitting procedure known as *generalised least squares* (GLS) can be used to provide better estimates of the standard errors of the regression parameters to account for the autocorrelation in the residual series. The procedure is essentially based on maximising the likelihood given the autocorrelation in the data and is implemented in R in the `gls` function (within the `nlme` library, which you will need to load).

5.4.1 GLS fit to simulated series

The following example illustrates how to fit a regression model to the simulated series of §5.2.3 using generalised least squares:

```
> library(nlme)
> x.gls <- gls(x ~ Time, cor = corAR1(0.8))
> coef(x.gls)

(Intercept)      Time
     58.23      3.04

> sqrt(diag(vcov(x.gls)))

(Intercept)      Time
    11.925     0.202
```

A lag 1 autocorrelation of 0.8 is used above because this value was used to simulate the data (§5.2.3). For historical series, the lag 1 autocorrelation would need to be estimated from the correlogram of the residuals of a fitted linear model; i.e., a linear model should first be fitted by ordinary least squares (OLS) and the lag 1 autocorrelation read off from a correlogram plot of the residuals of the fitted model.

In the example above, the standard errors of the parameters are considerably greater than those obtained from OLS using `lm` (§5.3) and are more accurate as they take the autocorrelation into account. The parameter estimates from GLS will generally be slightly different from those obtained with OLS, because of the weighting. For example, the slope is estimated as 3.06 using `lm` but 3.04 using `gls`. In principle, the GLS estimators are preferable because they have smaller standard errors.

5.4.2 Confidence interval for the trend in the temperature series

To calculate an approximate 95% confidence interval for the trend in the global temperature series (1970–2005), GLS is used to estimate the standard error accounting for the autocorrelation in the residual series (Fig. 5.4). In the `gls` function, the residual series is approximated as an AR(1) process with a lag 1 autocorrelation of 0.7 read from Figure 5.4, which is used as a parameter in the `gls` function:

```
> temp.gls <- gls(temp ~ time(temp), cor = corAR1(0.7))
> confint(temp.gls)
```

```
                2.5 %    97.5 %
(Intercept)  -39.8057  -28.4966
time(temp)     0.0144    0.0201
```

Although the confidence intervals above are now wider than they were in §5.3, zero is not contained in the intervals, which implies that the estimates are statistically significant, and, in particular, that the trend is significant. Thus, there is statistical evidence of an increasing trend in global temperatures over the period 1970–2005, so that, if current conditions persist, temperatures may be expected to continue to rise in the future.

5.5 Linear models with seasonal variables

5.5.1 Introduction

As time series are observations measured sequentially in time, seasonal effects are often present in the data, especially annual cycles caused directly or indirectly by the Earth's movement around the Sun. Seasonal effects have already been observed in several of the series we have looked at, including the airline series (§1.4.1), the temperature series (§1.4.5), and the electricity production series (§1.4.3). In this section, linear regression models with predictor variables for seasonal effects are considered.

5.5.2 Additive seasonal indicator variables

Suppose a time series contains s seasons. For example, with time series measured over each calendar month, $s = 12$, whereas for series measured over six-month intervals, corresponding to summer and winter, $s = 2$. A seasonal indicator model for a time series $\{x_t : t = 1, \ldots, n\}$ containing s seasons and a trend m_t is given by

$$x_t = m_t + s_t + z_t \tag{5.6}$$

where $s_t = \beta_i$ when t falls in the ith season ($t = 1, \ldots, n; i = 1, \ldots, s$) and $\{z_t\}$ is the residual error series, which may be autocorrelated. This model takes the same form as the additive decomposition model (Equation (1.2)) but differs in that the trend is formulated with parameters. In Equation (5.6), m_t does not have a constant term (referred to as the intercept), i.e., m_t could be a polynomial of order p with parameters $\alpha_1, \ldots, \alpha_p$. Equation (5.6) is then equivalent to a polynomial trend in which the constant term depends on the season, so that the s seasonal parameters $(\beta_1, \ldots, \beta_s)$ correspond to s possible constant terms in Equation (5.2). Equation (5.6) can therefore be written as

$$x_t = m_t + \beta_{1+(t-1)\bmod s} + z_t \tag{5.7}$$

For example, with a time series $\{x_t\}$ observed for each calendar month beginning with $t = 1$ at January, a seasonal indicator model with a straight-line trend is given by

$$x_t = \alpha_1 t + s_t + z_t = \begin{cases} \alpha_1 t + \beta_1 + z_t & t = 1, 13, \ldots \\ \alpha_1 t + \beta_2 + z_t & t = 2, 14, \ldots \\ \vdots & \\ \alpha_1 t + \beta_{12} + z_t & t = 12, 24, \ldots \end{cases} \tag{5.8}$$

The parameters for the model in Equation (5.8) can be estimated by OLS or GLS by treating the seasonal term s_t as a 'factor'. In R, the factor function can be applied to seasonal indices extracted using the function cycle (§1.4.1).

5.5.3 Example: Seasonal model for the temperature series

The parameters of a straight-line trend with additive seasonal indices can be estimated for the temperature series (1970–2005) as follows:

```
> Seas <- cycle(temp)
> Time <- time(temp)
> temp.lm <- lm(temp ~ 0 + Time + factor(Seas))
> coef(temp.lm)
```

```
           Time  factor(Seas)1  factor(Seas)2  factor(Seas)3
         0.0177       -34.9973       -34.9880       -35.0100
  factor(Seas)4  factor(Seas)5  factor(Seas)6  factor(Seas)7
       -35.0123       -35.0337       -35.0251       -35.0269
  factor(Seas)8  factor(Seas)9 factor(Seas)10 factor(Seas)11
       -35.0248       -35.0383       -35.0525       -35.0656
 factor(Seas)12
       -35.0487
```

A zero is used within the formula to ensure that the model does not have an intercept. If the intercept is included in the formula, one of the seasonal terms will be dropped and an estimate for the intercept will appear in the output. However, the fitted models, with or without an intercept, would be equivalent, as can be easily verified by rerunning the algorithm above without the zero in the formula. The parameters can also be estimated by GLS by replacing lm with gls in the code above.

Using the above fitted model, a two-year-ahead future prediction for the temperature series is obtained as follows:

```
> new.t <- seq(2006, len = 2 * 12, by = 1/12)
> alpha <- coef(temp.lm)[1]
> beta <- rep(coef(temp.lm)[2:13], 2)
> (alpha * new.t + beta)[1:4]
```

```
factor(Seas)1 factor(Seas)2 factor(Seas)3 factor(Seas)4
        0.524         0.535         0.514         0.514
```

Alternatively, the predict function can be used to make forecasts provided the new data are correctly labelled within a data.frame:

```
> new.dat <- data.frame(Time = new.t, Seas = rep(1:12, 2))
> predict(temp.lm, new.dat)[1:24]
```

```
     1     2     3     4     5     6     7     8     9    10    11    12
 0.524 0.535 0.514 0.514 0.494 0.504 0.503 0.507 0.495 0.482 0.471 0.489
    13    14    15    16    17    18    19    20    21    22    23    24
 0.542 0.553 0.532 0.531 0.511 0.521 0.521 0.525 0.513 0.500 0.488 0.507
```

5.6 Harmonic seasonal models

In the previous section, one parameter estimate is used per season. However, seasonal effects often vary smoothly over the seasons, so that it may be more parameter-efficient to use a smooth function instead of separate indices.

Sine and cosine functions can be used to build smooth variation into a seasonal model. A sine wave with frequency f (cycles per sampling interval), amplitude A, and phase shift ϕ can be expressed as

$$A\sin(2\pi ft + \phi) = \alpha_s \sin(2\pi ft) + \alpha_c \cos(2\pi ft) \qquad (5.9)$$

where $\alpha_s = A\cos(\phi)$ and $\alpha_c = A\sin(\phi)$. The expression on the right-hand side of Equation (5.9) is linear in the parameters α_s and α_c, whilst the left-hand side is non-linear because the parameter ϕ is within the sine function. Hence, the expression on the right-hand side is preferred in the formulation of a seasonal regression model, so that OLS can be used to estimate the parameters. For a time series $\{x_t\}$ with s seasons there are $[s/2]$ possible cycles.[1] The harmonic seasonal model is defined by

---

[1] The notation [ ] represents the integer part of the expression within. In most practical cases, s is even and so [ ] can be omitted. However, for some 'seasons', s may be an odd number, making the notation necessary. For example, if the 'seasons' are the days of the week, there would be $[7/2] = 3$ possible cycles.

$$x_t = m_t + \sum_{i=1}^{[s/2]} \left\{ s_i \sin(2\pi it/s) + c_i \cos(2\pi it/s) \right\} + z_t \qquad (5.10)$$

where m_t is the trend which includes a parameter for the constant term, and s_i and c_i are unknown parameters. The trend may take a polynomial form as in Equation (5.2). When s is an even number, the value of the sine at frequency $1/2$ (when $i = s/2$ in the summation term shown in Equation (5.10)) will be zero for all values of t, and so the term can be left out of the model. Hence, with a constant term included, the maximum number of parameters in the harmonic model equals that of the seasonal indicator variable model (Equation (5.6)), and the fits will be identical.

At first sight it may seem strange that the harmonic model has cycles of a frequency higher than the seasonal frequency of $1/s$. However, the addition of further harmonics has the effect of perturbing the underlying wave to make it less regular than a standard sine wave of period s. This usually still gives a dominant seasonal pattern of period s, but with a more realistic underlying shape. For example, suppose data are taken at monthly intervals. Then the second plot given below might be a more realistic underlying seasonal pattern than the first plot, as it perturbs the standard sine wave by adding another two harmonic terms of frequencies $2/12$ and $4/12$ (Fig. 5.5):

```
> TIME <- seq(1, 12, len = 1000)
> plot(TIME, sin(2 * pi * TIME/12), type = "l")
> plot(TIME, sin(2 * pi * TIME/12) + 0.2 * sin(2 * pi * 2 *
    TIME/12) + 0.1 * sin(2 * pi * 4 * TIME/12) + 0.1 *
    cos(2 * pi * 4 * TIME/12), type = "l")
```

The code above illustrates just one of many possible combinations of harmonics that could be used to model a wide range of possible underlying seasonal patterns.

5.6.1 Simulation

It is straightforward to simulate a series based on the harmonic model given by Equation (5.10). For example, suppose the underlying model is

$$
\begin{aligned}
x_t = {}& 0.1 + 0.005t + 0.001t^2 + \sin(2\pi t/12) + \\
& 0.2\sin(4\pi t/12) + 0.1\sin(8\pi t/12) + 0.1\cos(8\pi t/12) + w_t
\end{aligned}
\qquad (5.11)
$$

where $\{w_t\}$ is Gaussian white noise with standard deviation 0.5. This model has the same seasonal harmonic components as the model represented in Figure 5.5b but also contains an underlying quadratic trend. Using the code below, a series of length 10 years is simulated, and it is shown in Figure 5.6.

```
> set.seed(1)
> TIME <- 1:(10 * 12)
> w <- rnorm(10 * 12, sd = 0.5)
```

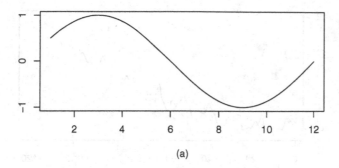

(a)

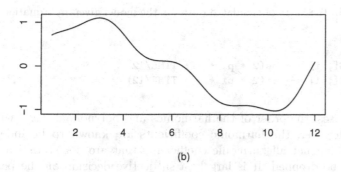

(b)

Fig. 5.5. Two possible underlying seasonal patterns for monthly series based on the harmonic model (Equation (5.10)). Plot (a) is of the first harmonic over a year and is usually too regular for most practical applications. Plot (b) is of the same wave but with a further two harmonics added. Plot (b) illustrates just one of many ways that an underlying sine wave can be perturbed to produce a less regular, but still dominant, seasonal pattern of period 12 months.

```
> Trend <- 0.1 + 0.005 * TIME + 0.001 * TIME^2
> Seasonal <- sin(2*pi*TIME/12) + 0.2*sin(2*pi*2*TIME/12) +
              0.1*sin(2*pi*4*TIME/12) + 0.1*cos(2*pi*4*TIME/12)
> x <- Trend + Seasonal + w
> plot(x, type = "l")
```

5.6.2 Fit to simulated series

With reference to Equation (5.10), it would seem reasonable to place the harmonic variables in matrices, which can be achieved as follows:

```
> SIN <- COS <- matrix(nr = length(TIME), nc = 6)
> for (i in 1:6) {
```

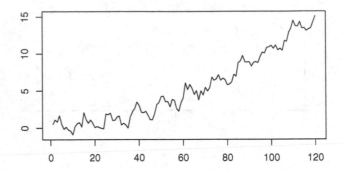

Fig. 5.6. Ten years of simulated data for the model given by Equation (5.11).

```
    COS[, i] <- cos(2 * pi * i * TIME/12)
    SIN[, i] <- sin(2 * pi * i * TIME/12)
}
```

In most cases, the order of the harmonics and polynomial trend will be unknown. However, the harmonic coefficients are known to be independent, which means that all harmonic coefficients that are not statistically significant can be dropped. It is largely a subjective decision on the part of the statistician to decide what constitutes a significant variable. An approximate t-ratio of magnitude 2 is a common choice and corresponds to an approximate 5% significance level. This t-ratio can be obtained by dividing the estimated coefficient by the standard error of the estimate. The following example illustrates the procedure applied to the simulated series of the last section:

```
> x.lm1 <- lm(x ~ TIME + I(TIME^2) + COS[, 1] + SIN[, 1] +
    COS[, 2] + SIN[, 2] + COS[, 3]  + SIN[, 3] + COS[, 4] +
    SIN[, 4] + COS[, 5] + SIN[, 5]  + COS[, 6] + SIN[, 6])
> coef(x.lm1)/sqrt(diag(vcov(x.lm1)))
```

(Intercept)	TIME	I(TIME^2)	COS[, 1]	SIN[, 1]	COS[, 2]
1.239	1.125	25.933	0.328	15.442	−0.515
SIN[, 2]	COS[, 3]	SIN[, 3]	COS[, 4]	SIN[, 4]	COS[, 5]
3.447	0.232	−0.703	0.228	1.053	−1.150
SIN[, 5]	COS[, 6]	SIN[, 6]			
0.857	−0.310	0.382			

The preceding output has three significant coefficients. These are used in the following model:[2]

---

[2] Some statisticians choose to include both the COS and SIN terms for a particular frequency if either has a statistically significant value.

```
> x.lm2 <- lm(x ~ I(TIME^2) + SIN[, 1] + SIN[, 2])
> coef(x.lm2)/sqrt(diag(vcov(x.lm2)))

(Intercept)   I(TIME^2)    SIN[, 1]     SIN[, 2]
       4.63      111.14       15.79         3.49
```

As can be seen in the output from the last command, the coefficients are all significant. The estimated coefficients of the best-fitting model are given by

```
> coef(x.lm2)

(Intercept)   I(TIME^2)    SIN[, 1]     SIN[, 2]
    0.28040     0.00104     0.90021      0.19886
```

The coefficients above give the following model for predictions at time t:

$$\hat{x}_t = 0.280 + 0.00104t^2 + 0.900\sin(2\pi t/12) + 0.199\sin(4\pi t/12) \qquad (5.12)$$

The AIC can be used to compare the two fitted models:

```
> AIC(x.lm1)

[1] 165

> AIC(x.lm2)

[1] 150
```

As expected, the last model has the smallest AIC and therefore provides the best fit to the data. Due to sampling variation, the best-fitting model is not identical to the model used to simulate the data, as can easily be verified by taking the AIC of the known underlying model:

```
> AIC(lm(x ~ TIME +I(TIME^2) +SIN[,1] +SIN[,2] +SIN[,4] +COS[,4]))

[1] 153
```

In R, the algorithm step can be used to automate the selection of the best-fitting model by the AIC. For the example above, the appropriate command is step(x.lm1), which contains all the predictor variables in the form of the first model. Try running this command, and check that the final output agrees with the model selected above.

A best fit can equally well be based on choosing the model that leads to the smallest estimated standard deviations of the errors, provided the degrees of freedom are taken into account.

5.6.3 Harmonic model fitted to temperature series (1970–2005)

In the code below, a harmonic model with a quadratic trend is fitted to the temperature series (1970–2005) from §5.3.2. The units for the 'time' variable are in 'years', so the divisor of 12 is not needed when creating the harmonic variables. To reduce computation error in the OLS procedure due to large numbers, the TIME variable is standardized after the COS and SIN predictors have been calculated.

```
> SIN <- COS <- matrix(nr = length(temp), nc = 6)
> for (i in 1:6) {
      COS[, i] <- cos(2 * pi * i * time(temp))
      SIN[, i] <- sin(2 * pi * i * time(temp))
 }
> TIME <- (time(temp) - mean(time(temp)))/sd(time(temp))
> mean(time(temp))

[1] 1988

> sd(time(temp))

[1] 10.4

> temp.lm1 <- lm(temp ~ TIME + I(TIME^2) +
                     COS[,1] + SIN[,1] + COS[,2] + SIN[,2] +
                     COS[,3] + SIN[,3] + COS[,4] + SIN[,4] +
                     COS[,5] + SIN[,5] + COS[,6] + SIN[,6])
> coef(temp.lm1)/sqrt(diag(vcov(temp.lm1)))
```

(Intercept)	TIME	I(TIME^2)	COS[, 1]	SIN[, 1]	COS[, 2]
18.245	30.271	1.281	0.747	2.383	1.260
SIN[, 2]	COS[, 3]	SIN[, 3]	COS[, 4]	SIN[, 4]	COS[, 5]
1.919	0.640	0.391	0.551	0.168	0.324
SIN[, 5]	COS[, 6]	SIN[, 6]			
0.345	-0.409	-0.457			

```
> temp.lm2 <- lm(temp ~ TIME + SIN[, 1] + SIN[, 2])
> coef(temp.lm2)
```

(Intercept)	TIME	SIN[, 1]	SIN[, 2]
0.1750	0.1841	0.0204	0.0162

```
> AIC(temp.lm)

[1] -547

> AIC(temp.lm1)

[1] -545

> AIC(temp.lm2)

[1] -561
```

Again, the AIC is used to compare the fitted models, and only statistically significant terms are included in the final model.

To check the adequacy of the fitted model, it is appropriate to create a time plot and correlogram of the residuals because the residuals form a time series (Fig. 5.7). The time plot is used to detect patterns in the series. For example, if a higher-ordered polynomial is required, this would show up as a curve in the time plot. The purpose of the correlogram is to determine whether there is autocorrelation in the series, which would require a further model.

```
> plot(time(temp), resid(temp.lm2), type = "l")
> abline(0, 0, col = "red")
> acf(resid(temp.lm2))
> pacf(resid(temp.lm2))
```

In Figure 5.7(a), there is no discernible curve in the series, which implies that a straight line is an adequate description of the trend. A tendency for the series to persist above or below the x-axis implies that the series is positively autocorrelated. This is verified in the correlogram of the residuals, which shows a clear positive autocorrelation at lags 1–10 (Fig. 5.7b).

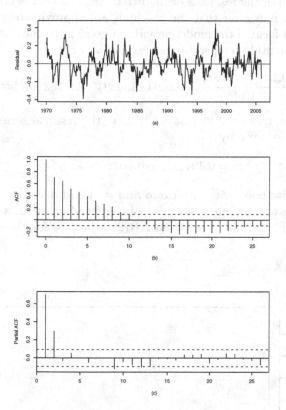

Fig. 5.7. Residual diagnostic plots for the harmonic model fitted to the temperature series (1970–2005): (a) the residuals plotted against time; (b) the correlogram of the residuals (time units are months); (c) partial autocorrelations plotted against lag (in months).

The correlogram in Figure 5.7 is similar to that expected of an AR(p) process (§4.5.5). This is verified by the plot of the partial autocorrelations, in which only the lag 1 and lag 2 autocorrelations are statistically significant (Fig. 5.7). In the code below, an AR(2) model is fitted to the residual series:

```
> res.ar <- ar(resid(temp.lm2), method = "mle")
> res.ar$ar
```

```
[1] 0.494 0.307
```

```
> sd(res.ar$res[-(1:2)])
```

```
[1] 0.0837
```

```
> acf(res.ar$res[-(1:2)])
```

The correlogram of the residuals of the fitted AR(2) model is given in Figure 5.8, from which it is clear that the residuals are approximately white noise. Hence, the final form of the model provides a good fit to the data. The fitted model for the monthly temperature series can be written as

$$x_t = 0.175 + \frac{0.184(t - 1988)}{10.4} + 0.0204\sin(2\pi t) + 0.0162\sin(4\pi t) + z_t \quad (5.13)$$

where t is 'time' measured in units of 'years', the residual series $\{z_t\}$ follow an AR(2) process given by

$$z_t = 0.494z_{t-1} + 0.307z_{t-2} + w_t \quad (5.14)$$

and $\{w_t\}$ is white noise with mean zero and standard deviation 0.0837.

If we require an accurate assessment of the standard error, we should refit the model using gls, allowing for an AR(2) structure for the errors (Exercise 6).

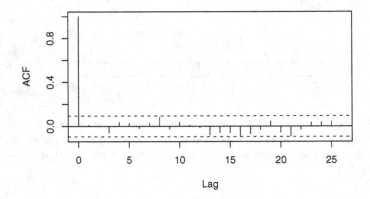

Fig. 5.8. Correlogram of the residuals of the AR(2) model fitted to the residuals of the harmonic model for the temperature series.

5.7 Logarithmic transformations

5.7.1 Introduction

Recall from §5.2 that the natural logarithm (base e) can be used to transform a model with multiplicative components to a model with additive components. For example, if $\{x_t\}$ is a time series given by

$$x_t = m_t' \, s_t' \, z_t' \qquad (5.15)$$

where m_t' is the trend, s_t' is the seasonal effect, and z_t' is the residual error, then the series $\{y_t\}$, given by

$$y_t = \log x_t = \log m_t' + \log s_t' + \log z_t' = m_t + s_t + z_t \qquad (5.16)$$

has additive components, so that if m_t and s_t are also linear functions, the parameters in Equation (5.16) can be estimated by OLS. In Equation (5.16), logs can be taken only if the series $\{x_t\}$ takes all positive values; i.e., $x_t > 0$ for all t. Conversely, a log-transformation may be seen as an appropriate model formulation when a series can only take positive values and has values near zero because the anti-log forces the predicted and simulated values for $\{x_t\}$ to be positive.

5.7.2 Example using the air passenger series

Consider the air passenger series from §1.4.1. Time plots of the original series and the natural logarithm of the series can be obtained using the code below and are shown in Figure 5.9.

```
> data(AirPassengers)
> AP <- AirPassengers
> plot(AP)
> plot(log(AP))
```

In Figure 5.9(a), the variance can be seen to increase as t increases, whilst after the logarithm is taken the variance is approximately constant over the period of the record (Fig. 5.9b). Therefore, as the number of people using the airline can also only be positive, the logarithm would be appropriate in the model formulation for this time series. In the following code, a harmonic model with polynomial trend is fitted to the air passenger series. The function time is used to extract the time and create a standardised time variable TIME.

```
> SIN <- COS <- matrix(nr = length(AP), nc = 6)
> for (i in 1:6) {
    SIN[, i] <- sin(2 * pi * i * time(AP))
    COS[, i] <- cos(2 * pi * i * time(AP))
  }
> TIME <- (time(AP) - mean(time(AP)))/sd(time(AP))
> mean(time(AP))
```

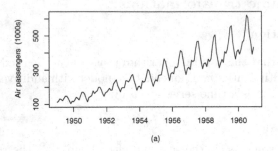

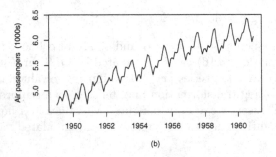

Fig. 5.9. Time plots of (a) the airline series (1949–1960) and (b) the natural logarithm of the airline series.

```
[1] 1955

> sd(time(AP))

[1] 3.48

> AP.lm1 <- lm(log(AP) ~ TIME + I(TIME^2) + I(TIME^3) + I(TIME^4) +
        SIN[,1] + COS[,1] + SIN[,2] + COS[,2] + SIN[,3] + COS[,3] +
        SIN[,4] + COS[,4] + SIN[,5] + COS[,5] + SIN[,6] + COS[,6])
> coef(AP.lm1)/sqrt(diag(vcov(AP.lm1)))

(Intercept)       TIME  I(TIME^2)  I(TIME^3)  I(TIME^4)   SIN[, 1]
    744.685     42.382     -4.162     -0.751      1.873      4.868
   COS[, 1]   SIN[, 2]   COS[, 2]   SIN[, 3]   COS[, 3]   SIN[, 4]
    -26.055     10.395     10.004     -4.844     -1.560     -5.666
   COS[, 4]   SIN[, 5]   COS[, 5]   SIN[, 6]   COS[, 6]
      1.946     -3.766      1.026      0.150     -0.521

> AP.lm2 <- lm(log(AP) ~ TIME + I(TIME^2) + SIN[,1] + COS[,1] +
    SIN[,2] + COS[,2] + SIN[,3] + SIN[,4] + COS[,4] + SIN[,5])
> coef(AP.lm2)/sqrt(diag(vcov(AP.lm2)))
```

(Intercept)	TIME	I(TIME^2)	SIN[, 1]	COS[, 1]	SIN[, 2]
922.63	103.52	-8.24	4.92	-25.81	10.36
COS[, 2]	SIN[, 3]	SIN[, 4]	COS[, 4]	SIN[, 5]	
9.96	-4.79	-5.61	1.95	-3.73	

```
> AIC(AP.lm1)

[1] -448

> AIC(AP.lm2)

[1] -451

> acf(resid(AP.lm2))
```

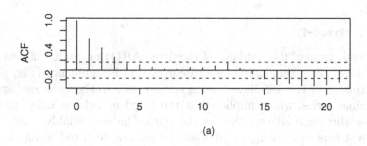

(a)

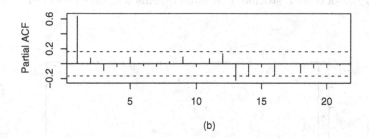

(b)

Fig. 5.10. The correlogram (a) and partial autocorrelations (b) of the residual series.

The residual correlogram indicates that the data are positively autocorrelated (Fig. 5.10). As mentioned in §5.4, the standard errors of the parameter estimates are likely to be under-estimated if there is positive serial correlation in the data. This implies that predictor variables may falsely appear 'significant' in the fitted model. In the code below, GLS is used to check the significance of the variables in the fitted model, using the lag 1 autocorrelation (approximately 0.6) from Figure 5.10.

```
> AP.gls <- gls(log(AP) ~ TIME + I(TIME^2) + SIN[,1] + COS[,1] +
    SIN[,2] + COS[,2] + SIN[,3] + SIN[,4] + COS[,4] + SIN[,5],
    cor = corAR1(0.6))
> coef(AP.gls)/sqrt(diag(vcov(AP.gls)))
```

```
(Intercept)       TIME   I(TIME^2)    SIN[, 1]   COS[, 1]    SIN[, 2]
     398.84      45.85       -3.65        3.30     -18.18       11.77
   COS[, 2]   SIN[, 3]    SIN[, 4]    COS[, 4]   SIN[, 5]
      11.43      -7.63      -10.75        3.57      -7.92
```

In Figure 5.10(b), the partial autocorrelation plot suggests that the residual series follows an AR(1) process, which is fitted to the series below:

```
> AP.ar <- ar(resid(AP.lm2), order = 1, method = "mle")
> AP.ar$ar
```

```
[1] 0.641
```

```
> acf(AP.ar$res[-1])
```

The correlogram of the residuals of the fitted AR(1) model might be taken for white noise given that only one autocorrelation is significant (Fig. 5.11). However, the lag of this significant value corresponds to the seasonal lag (12) in the original series, which implies that the fitted model has failed to fully account for the seasonal variation in the data. Understandably, the reader might regard this as curious, given that the data were fitted using the full seasonal harmonic model. However, seasonal effects can be stochastic just as trends can, and the harmonic model we have used is deterministic. In Chapter 7, models with stochastic seasonal terms will be considered.

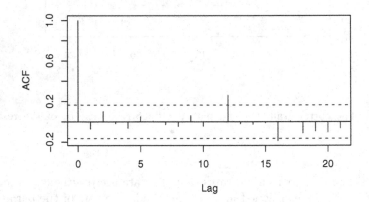

Fig. 5.11. Correlogram of the residuals from the AR(1) model fitted to the residuals of the logarithm model.

5.8 Non-linear models

5.8.1 Introduction

For the reasons given in §5.2, linear models are applicable to a wide range of time series. However, for some time series it may be more appropriate to fit a non-linear model directly rather than take logs or use a linear polynomial approximation. For example, if a series is known to derive from a known non-linear process, perhaps based on an underlying known deterministic law in science, then it would be better to use this information in the model formulation and fit a non-linear model directly to the data. In R, a non-linear model can be fitted by least squares using the function nls.

In the previous section, we found that using the natural logarithm of a series could help stabilise the variance. However, using logs can present difficulties when a series contains negative values, because the log of a negative value is undefined. One way around this problem is to add a constant to all the terms in the series, so if $\{x_t\}$ is a series containing (some) negative values, then adding c_0 such that $c_0 > \max\{-x_t\}$ and then taking logs produces a transformed series $\{\log(c_0 + x_t)\}$ that is defined for all t. A linear model (e.g., a straight-line trend) could then be fitted to produce for $\{x_t\}$ the model

$$x_t = -c_0 + e^{\alpha_0 + \alpha_1 t + z_t} \tag{5.17}$$

where α_0 and α_1 are model parameters and $\{z_t\}$ is a residual series that may be autocorrelated.

The main difficulty with the approach leading to Equation (5.17) is that c_0 should really be estimated like any other parameter in the model, whilst in practice a user will often arbitrarily choose a value that satisfies the constraint $(c_0 > \max\{-x_t\})$. If there is a reason to expect a model similar to that in Equation (5.17) but there is no evidence for multiplicative residual terms, then the constant c_0 should be estimated with the other model parameters using non-linear least squares; i.e., the following model should be fitted:

$$x_t = -c_0 + e^{\alpha_0 + \alpha_1 t} + z_t \tag{5.18}$$

5.8.2 Example of a simulated and fitted non-linear series

As non-linear models are generally fitted when the underlying non-linear function is known, we will simulate a non-linear series based on Equation (5.18) with $c_0 = 0$ and compare parameters estimated using nls with those of the known underlying function.

Below, a non-linear series with AR(1) residuals is simulated and plotted (Fig. 5.12):

```
> set.seed(1)
> w <- rnorm(100, sd = 10)
```

```
> z <- rep(0, 100)
> for (t in 2:100) z[t] <- 0.7 * z[t - 1] + w[t]
> Time <- 1:100
> f <- function(x) exp(1 + 0.05 * x)
> x <- f(Time) + z
> plot(x, type = "l")
> abline(0, 0)
```

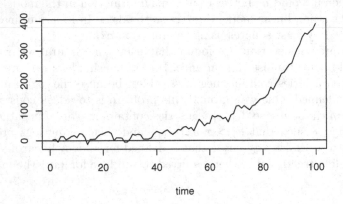

Fig. 5.12. Plot of a non-linear series containing negative values.

The series plotted in Figure 5.12 has an apparent increasing exponential trend but also contains negative values, so that a direct log-transformation cannot be used and a non-linear model is needed. In R, a non-linear model is fitted by specifying a formula with the parameters and their starting values contained in a `list`:

```
> x.nls <- nls(x ~ exp(alp0 + alp1 * Time), start = list(alp0 = 0.1,
    alp1 = 0.5))
> summary(x.nls)$parameters
```

	Estimate	Std. Error	t value	Pr(>|t|)
alp0	1.1764	0.074295	15.8	9.20e-29
alp1	0.0483	0.000819	59.0	2.35e-78

The estimates for α_0 and α_1 are close to the underlying values that were used to simulate the data, although the standard errors of these estimates are likely to be underestimated because of the autocorrelation in the residuals.[3]

---

[3] The generalised least squares function `gls` can be used to fit non-linear models with autocorrelated residuals. However, in practice, computational difficulties often arise when using this function with non-linear models.

5.9 Forecasting from regression

5.9.1 Introduction

A forecast is a prediction into the future. In the context of time series regression, a forecast involves extrapolating a fitted model into the future by evaluating the model function for a new series of times. The main problem with this approach is that the trends present in the fitted series may change in the future. Therefore, it is better to think of a forecast from a regression model as an expected value conditional on past trends continuing into the future.

5.9.2 Prediction in R

The generic function for making predictions in R is `predict`. The function essentially takes a fitted model and new data as parameters. The key to using this function with a regression model is to ensure that the new data are properly defined and labelled in a `data.frame`.

In the code below, we use this function in the fitted regression model of §5.7.2 to forecast the number of air passengers travelling for the 10-year period that follows the record (Fig. 5.13). The forecast is given by applying the exponential function (anti-log) to `predict` because the regression model was fitted to the logarithm of the series:

```
> new.t <- time(ts(start = 1961, end = c(1970, 12), fr = 12))
> TIME <- (new.t - mean(time(AP)))/sd(time(AP))
> SIN <- COS <- matrix(nr = length(new.t), nc = 6)
> for (i in 1:6) {
      COS[, i] <- cos(2 * pi * i * new.t)
      SIN[, i] <- sin(2 * pi * i * new.t)
  }
> SIN <- SIN[, -6]
> new.dat <- data.frame(TIME = as.vector(TIME), SIN = SIN,
      COS = COS)
> AP.pred.ts <- exp(ts(predict(AP.lm2, new.dat), st = 1961,
      fr = 12))
> ts.plot(log(AP), log(AP.pred.ts), lty = 1:2)
> ts.plot(AP, AP.pred.ts, lty = 1:2)
```

5.10 Inverse transform and bias correction

5.10.1 Log-normal residual errors

The forecasts in Figure 5.13(b) were obtained by applying the anti-log to the forecasted values obtained from the log-regression model. However, the process

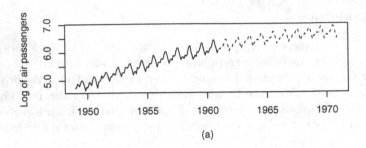

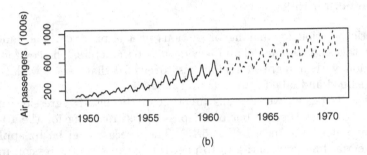

Fig. 5.13. Air passengers (1949–1960; solid line) and forecasts (1961–1970; dotted lines): (a) logarithm and forecasted values; (b) original series and anti-log of the forecasted values.

of using a transformation, such as the logarithm, and then applying an inverse transformation introduces a bias in the forecasts of the mean values. If the regression model closely fits the data, this bias will be small (as shown in the next example for the airline predictions). Note that a bias correction is only for means and should not be used in simulations.

The bias in the means arises as a result of applying the inverse transform to a residual series. For example, if the time series are Gaussian white noise $\{w_t\}$, with mean zero and standard deviation σ, then the distribution of the inverse-transform (the anti-log) of the series is log-normal with mean $e^{\frac{1}{2}\sigma^2}$. This can be verified theoretically, or empirically by simulation as in the code below:

```
> set.seed(1)
> sigma <- 1
> w <- rnorm(1e+06, sd = sigma)
> mean(w)

[1] 4.69e-05
```

```
> mean(exp(w))
```

```
[1] 1.65
```

```
> exp(sigma^2/2)
```

```
[1] 1.65
```

The code above indicates that the mean of the anti-log of the Gaussian white noise and the expected mean from a log-normal distribution are equal. Hence, for a Gaussian white noise residual series, a correction factor of $e^{\frac{1}{2}\sigma^2}$ should be applied to the forecasts of means. The importance of this correction factor really depends on the value of σ^2. If σ^2 is very small, the correction factor will hardly change the forecasts at all and so could be neglected without major concern, especially as errors from other sources are likely to be significantly greater.

5.10.2 Empirical correction factor for forecasting means

The $e^{\frac{1}{2}\sigma^2}$ correction factor can be used when the residual series of the fitted log-regression model is Gaussian white noise. In general, however, the distribution of the residuals from the log regression (Exercise 5) is often negatively skewed, in which case a correction factor can be determined empirically using the mean of the anti-log of the residual series. In this approach, adjusted forecasts $\{\hat{x}'_t\}$ can be obtained from

$$\hat{x}'_t = e^{\hat{\log} x_t} \sum_{t=1}^{n} e^{z_t}/n \tag{5.19}$$

where $\{\hat{\log} x_t : t = 1, \ldots, n\}$ is the predicted series given by the fitted log-regression model, and $\{z_t\}$ is the residual series from this fitted model.

The following example illustrates the procedure for calculating the correction factors.

5.10.3 Example using the air passenger data

For the airline series, the forecasts can be adjusted by multiplying the predictions by $e^{\frac{1}{2}\sigma^2}$, where σ is the standard deviation of the residuals, or using an empirical correction factor as follows:

```
> summary(AP.lm2)$r.sq
```

```
[1] 0.989
```

```
> sigma <- summary(AP.lm2)$sigma
> lognorm.correction.factor <- exp((1/2) * sigma^2)
> empirical.correction.factor <- mean(exp(resid(AP.lm2)))
```

```
> lognorm.correction.factor
```

```
[1] 1.001171
```

```
> empirical.correction.factor
```

```
[1] 1.001080
```

```
> AP.pred.ts <- AP.pred.ts * empirical.correction.factor
```

The adjusted forecasts in AP.pred.ts allow for the bias in taking the anti-log of the predictions. However, the small σ (and $R^2 = 0.99$) results in a small correction factor (of the order 0.1%), which is probably negligible compared with other sources of errors that exist in the forecasts. Whilst in this example the correction factor is small, there is no reason why it will be small in general.

5.11 Summary of R commands

lm	fits a linear (regression) model
coef	extracts the parameter estimates from a fitted model
confint	returns a (95%) confidence interval for the parameters of a fitted model
gls	fits a linear model using generalised least squares (allowing for autocorrelated residuals)
factor	returns variables in the form of 'factors' or indicator variables

5.12 Exercises

1. a) Produce a time plot for $\{x_t : t = 1, \ldots, 100\}$, where $x_t = 70 + 2t - 3t^2 + z_t$, $\{z_t\}$ is the AR(1) process $z_t = 0.5z_{t-1} + w_t$, and $\{w_t\}$ is white noise with standard deviation 25.

   b) Fit a quadratic trend to the series $\{x_t\}$. Give the coefficients of the fitted model.

   c) Find a 95% confidence interval for the parameters of the quadratic model, and comment.

   d) Plot the correlogram of the residuals and comment.

   e) Refit the model using GLS. Give the standard errors of the parameter estimates, and comment.

2. The standard errors of the parameter estimates of a fitted regression model are likely to be underestimated if there is positive serial correlation in the data. This implies that explanatory variables may appear as 'significant' when they should not. Use GLS to check the significance of the variables

of the fitted model from §5.6.3. Use an appropriate estimate of the lag 1 autocorrelation within `gls`.

3. This question is based on the electricity production series (1958–1990).
    a) Give two reasons why a log-transformation may be appropriate for the electricity series.
    b) Fit a seasonal indicator model with a quadratic trend to the (natural) logarithm of the series. Use stepwise regression to select the best model based on the AIC.
    c) Fit a harmonic model with a quadratic trend to the logarithm of the series. Use stepwise regression to select the best model based on the AIC.
    d) Plot the correlogram and partial correlogram of the residuals from the overall best-fitting model and comment on the plots.
    e) Fit an AR model to the residuals of the best-fitting model. Give the order of the best-fitting AR model and the estimated model parameters.
    f) Plot the correlogram of the residuals of the AR model, and comment.
    g) Write down in full the equation of the best-fitting model.
    h) Use the best fitting model to forecast electricity production for the years 1991–2000, making sure you have corrected for any bias due to taking logs.

4. Suppose a sample of size n follows an AR(1) process with lag 1 autocorrelation $\rho_1 = \alpha$. Use Equation (5.5) to find the variance of the sample mean.

5. A hydrologist wishes to simulate monthly inflows to the Font Reservoir over the next 10-year period. Use the data in `Font.dat` (§2.3.3) to answer the following:
    a) Regress `inflow` on `month` using indicator variables and time t, and fit a suitable AR model to the residual error series.
    b) Plot a histogram of the residual errors of the fitted AR model, and comment on the plot. Fit back-to-back Weibull distributions to the errors.
    c) Simulate 20 realisations of `inflow` for the next 10 years.
    d) Give reasons why a log transformation may be suitable for the series of inflows.
    e) Regress `log(inflow)` on `month` using indicator variables and time t (as above), and fit a suitable AR model to the residual error series.
    f) Plot a histogram of the residual errors of the fitted AR model, and comment on the plot. Fit a back-to-back Weibull distribution to the residual errors.

g) Simulate 20 realisations of `log(inflow)` for the next 10-years. Take anti-logs of the simulated values to produce a series of simulated flows.

h) Compare both sets of simulated flows, and discuss which is the more satisfactory.

6. Refit the harmonic model to the temperature series using `gls`, allowing for errors from an AR(2) process.

a) Construct a 99% confidence interval for the coefficient of time.

b) Plot the residual error series from the model fitted using GLS against the residual error series from the model fitted using OLS.

c) Refit the AR(2) model to the residuals from the fitted (GLS) model.

d) How different are the fitted models?

e) Calculate the annual means. Use OLS to regress the annual mean temperature on time, and construct a 99% confidence interval for its coefficient.

6

Stationary Models

6.1 Purpose

As seen in the previous chapters, a time series will often have well-defined components, such as a trend and a seasonal pattern. A well-chosen linear regression may account for these non-stationary components, in which case the residuals from the fitted model should not contain noticeable trend or seasonal patterns. However, the residuals will usually be correlated in time, as this is not accounted for in the fitted regression model. Similar values may cluster together in time; for example, monthly values of the Southern Oscillation Index, which is closely associated with El Niño, tend to change slowly and may give rise to persistent weather patterns. Alternatively, adjacent observations may be negatively correlated; for example, an unusually high monthly sales figure may be followed by an unusually low value because customers have supplies left over from the previous month. In this chapter, we consider stationary models that may be suitable for residual series that contain no obvious trends or seasonal cycles. The fitted stationary models may then be combined with the fitted regression model to improve forecasts. The autoregressive models that were introduced in §4.5 often provide satisfactory models for the residual time series, and we extend the repertoire in this chapter. The term *stationary* was discussed in previous chapters; we now give a more rigorous definition.

6.2 Strictly stationary series

A time series model $\{x_t\}$ is *strictly stationary* if the joint statistical distribution of $x_{t_1}, \ldots, x_{t_n}$ is the same as the joint distribution of $x_{t_1+m}, \ldots, x_{t_n+m}$ for all $t_1, \ldots, t_n$ and m, so that the distribution is unchanged after an arbitrary time shift. Note that strict stationarity implies that the mean and variance are constant in time and that the autocovariance $\text{Cov}(x_t, x_s)$ only depends on lag $k = |t - s|$ and can be written $\gamma(k)$. If a series is not strictly stationary but the mean and variance are constant in time and the autocovariance only

P.S.P. Cowpertwait and A.V. Metcalfe, *Introductory Time Series with R*,   121
Use R, DOI 10.1007/978-0-387-88698-5_6,
© Springer Science+Business Media, LLC 2009

depends on the lag, then the series is called *second-order* stationary.[1] We focus on the second-order properties in this chapter, but the stochastic processes discussed are strictly stationary. Furthermore, if the white noise is Gaussian, the stochastic process is completely defined by the mean and covariance structure, in the same way as any normal distribution is defined by its mean and variance-covariance matrix.

Stationarity is an idealisation that is a property of models. If we fit a stationary model to data, we assume our data are a realisation of a stationary process. So our first step in an analysis should be to check whether there is any evidence of a trend or seasonal effects and, if there is, remove them. Regression can break down a non-stationary series to a trend, seasonal components, and residual series. It is often reasonable to treat the time series of residuals as a realisation of a stationary error series. Therefore, the models in this chapter are often fitted to residual series arising from regression analyses.

6.3 Moving average models

6.3.1 MA(q) process: Definition and properties

A moving average (MA) process of order q is a linear combination of the current white noise term and the q most recent past white noise terms and is defined by

$$x_t = w_t + \beta_1 w_{t-1} + \ldots + \beta_q w_{t-q} \tag{6.1}$$

where $\{w_t\}$ is white noise with zero mean and variance σ_w^2. Equation (6.1) can be rewritten in terms of the backward shift operator $\mathbf{B}$

$$x_t = (1 + \beta_1 \mathbf{B} + \beta_2 \mathbf{B}^2 + \cdots + \beta_q \mathbf{B}^q) w_t = \phi_q(\mathbf{B}) w_t \tag{6.2}$$

where ϕ_q is a polynomial of order q. Because MA processes consist of a finite sum of stationary white noise terms, they are stationary and hence have a time-invariant mean and autocovariance.

The mean and variance for $\{x_t\}$ are easy to derive. The mean is just zero because it is a sum of terms that all have a mean of zero. The variance is $\sigma_w^2(1 + \beta_1^2 + \ldots + \beta_q^2)$ because each of the white noise terms has the same variance and the terms are mutually independent. The autocorrelation function, for $k \geq 0$, is given by

$$\rho(k) = \begin{cases} 1 & k = 0 \\ \sum_{i=0}^{q-k} \beta_i \beta_{i+k} / \sum_{i=0}^{q} \beta_i^2 & k = 1, \ldots, q \\ 0 & k > q \end{cases} \tag{6.3}$$

where β_0 is unity. The function is zero when $k > q$ because x_t and x_{t+k} then consist of sums of independent white noise terms and so have covariance

---

[1] For example, the skewness, or more generally $\mathrm{E}(x_t x_{t+k} x_{t+l})$, might change over time.

zero. The derivation of the autocorrelation function is left to Exercise 1. An MA process is *invertible* if it can be expressed as a stationary autoregressive process of infinite order without an error term. For example, the MA process $x_t = (1 - \beta\mathbf{B})w_t$ can be expressed as

$$w_t = (1 - \beta\mathbf{B})^{-1}x_t = x_t + \beta x_{t-1} + \beta^2 x_{t-2} + \dots \tag{6.4}$$

provided $|\beta| < 1$, which is required for convergence.

In general, an MA(q) process is invertible when the roots of $\phi_q(B)$ all exceed unity in absolute value (Exercise 2). The autocovariance function only identifies a unique MA(q) process if the condition that the process be invertible is imposed. The estimation procedure described in §6.4 leads naturally to invertible models.

6.3.2 R examples: Correlogram and simulation

The autocorrelation function for an MA(q) process (Equation (6.3)) can readily be implemented in R, and a simple version, without any detailed error checks, is given below. Note that the function takes the lag k and the model parameters β_i for $i = 0, 1, \dots, q$, with $\beta_0 = 1$. For the non-zero values (i.e., values within the `else` part of the `if-else` statement), the autocorrelation function is computed in two stages using a `for` loop. The first loop generates a sum (`s1`) for the autocovariance, whilst the second loop generates a sum (`s2`) for the variance, with the division of the two sums giving the returned autocorrelation (`ACF`).

```
> rho <- function(k, beta) {
    q <- length(beta) - 1
    if (k > q) ACF <- 0 else {
      s1 <- 0; s2 <- 0
      for (i in 1:(q-k+1)) s1 <- s1 + beta[i] * beta[i+k]
      for (i in 1:(q+1)) s2 <- s2 + beta[i]^2
      ACF <- s1 / s2}
    ACF}
```

Using the code above for the autocorrelation function, correlograms for a range of MA(q) processes can be plotted against lag – the code below provides an example for an MA(3) process with parameters $\beta_1 = 0.7$, $\beta_2 = 0.5$, and $\beta_3 = 0.2$ (Fig. 6.1a).

```
> beta <- c(1, 0.7, 0.5, 0.2)
> rho.k <- rep(1, 10)
> for (k in 1:10) rho.k[k] <- rho(k, beta)
> plot(0:10, c(1, rho.k), pch = 4, ylab = expression(rho[k]))
> abline(0, 0)
```

The plot in Figure 6.1(b) is the autocovariance function for an MA(3) process with parameters $\beta_1 = -0.7$, $\beta_2 = 0.5$, and $\beta_3 = -0.2$, which has negative

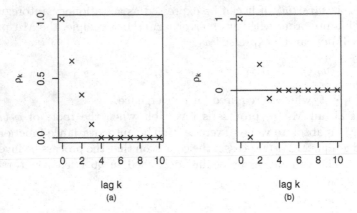

Fig. 6.1. Plots of the autocorrelation functions for two MA(3) processes: (a) $\beta_1 = 0.7$, $\beta_2 = 0.5$, $\beta_3 = 0.2$; (b) $\beta_1 = -0.7$, $\beta_2 = 0.5$, $\beta_3 = -0.2$.

correlations at lags 1 and 3. The function **expression** is used to get the Greek symbol ρ.

The code below can be used to simulate the MA(3) process and plot the correlogram of the simulated series. An example time plot and correlogram are shown in Figure 6.2. As expected, the first three autocorrelations are significantly different from 0 (Fig. 6.2b); other statistically significant correlations are attributable to random sampling variation. Note that in the correlogram plot (Fig. 6.2b) 1 in 20 (5%) of the sample correlations for lags greater than 3, for which the underlying population correlation is zero, are expected to be statistically significantly different from zero at the 5% level because multiple t-test results are being shown on the plot.

```
> set.seed(1)
> b <- c(0.8, 0.6, 0.4)
> x <- w <- rnorm(1000)
> for (t in 4:1000) {
      for (j in 1:3) x[t] <- x[t] + b[j] * w[t - j]
  }
> plot(x, type = "l")
> acf(x)
```

6.4 Fitted MA models

6.4.1 Model fitted to simulated series

An MA(q) model can be fitted to data in R using the **arima** function with the **order** function parameter set to c(0,0,q). Unlike the function **ar**, the

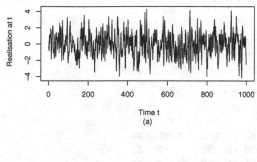

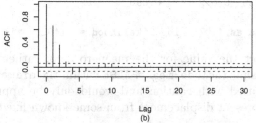

Fig. 6.2. (a) Time plot and (b) correlogram for a simulated MA(3) process.

function `arima` does not subtract the mean by default and estimates an intercept term. MA models cannot be expressed in a multiple regression form, and, in general, the parameters are estimated with a numerical algorithm. The function `arima` minimises the conditional sum of squares to estimate values of the parameters and will either return these if `method=c("CSS")` is specified or use them as initial values for maximum likelihood estimation.

A description of the conditional sum of squares algorithm for fitting an MA(q) process follows. For any choice of parameters, the sum of squared residuals can be calculated iteratively by rearranging Equation (6.1) and replacing the errors, w_t, with their estimates (that is, the residuals), which are denoted by $\hat{w}_t$:

$$S(\hat{\beta}_1, \ldots, \hat{\beta}_q) = \sum_{t=1}^{n} \hat{w}_t^2 = \sum_{t=1}^{n} \left\{ x_t - (\hat{\beta}_1 \hat{w}_{t-1} + \cdots + \hat{\beta}_q \hat{w}_{t-q}) \right\}^2 \qquad (6.5)$$

conditional on $\hat{w}_0, \ldots, \hat{w}_{t-q}$ being taken as 0 to start the iteration. A numerical search is used to find the parameter values that minimise this conditional sum of squares.

In the following code, a moving average model, `x.ma`, is fitted to the simulated series of the last section. Looking at the parameter estimates (coefficients in the output below), it can be seen that the 95% confidence intervals (approximated by *coeff.* ±2 s.e. of *coeff.*) contain the underlying parameter values (0.8, 0.6, and 0.4) that were used in the simulations. Furthermore, also as expected, the intercept is not significantly different from its underlying parameter value of zero.

```
> x.ma <- arima(x, order = c(0, 0, 3))
> x.ma

Call:
arima(x = x, order = c(0, 0, 3))

Coefficients:
         ma1    ma2    ma3  intercept
       0.790  0.566  0.396    -0.032
s.e.   0.031  0.035  0.032     0.090

sigma^2 estimated as 1.07:  log likelihood = -1452,  aic = 2915
```

It is possible to set the value for the mean to zero, rather than estimate the intercept, by using include.mean=FALSE within the arima function. This option should be used with caution and would only be appropriate if you wanted $\{x_t\}$ to represent displacement from some known fixed mean.

6.4.2 Exchange rate series: Fitted MA model

In the code below, an MA(1) model is fitted to the exchange rate series. If you refer back to §4.6.2, a comparison with the output below indicates that the AR(1) model provides the better fit, as it has the smaller standard deviation of the residual series, 0.031 compared with 0.042. Furthermore, the correlogram of the residuals indicates that an MA(1) model does not provide a satisfactory fit, as the residual series is clearly not a realistic realisation of white noise (Fig. 6.3).

```
> www <- "http://www.massey.ac.nz/~pscowper/ts/pounds_nz.dat"
> x <- read.table(www, header = T)
> x.ts <- ts(x, st = 1991, fr = 4)
> x.ma <- arima(x.ts, order = c(0, 0, 1))
> x.ma

Call:
arima(x = x.ts, order = c(0, 0, 1))

Coefficients:
         ma1  intercept
       1.000      2.833
s.e.   0.072      0.065

sigma^2 estimated as 0.0417:  log likelihood = 4.76,  aic = -3.53

> acf(x.ma$res[-1])
```

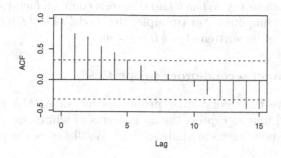

Fig. 6.3. The correlogram of residual series for the MA(1) model fitted to the exchange rate data.

6.5 Mixed models: The ARMA process

6.5.1 Definition

Recall from Chapter 4 that a series $\{x_t\}$ is an autoregressive process of order p, an AR(p) process, if

$$x_t = \alpha_1 x_{t-1} + \alpha_2 x_{t-2} + \ldots + \alpha_p x_{t-p} + w_t \qquad (6.6)$$

where $\{w_t\}$ is white noise and the α_i are the model parameters. A useful class of models are obtained when AR and MA terms are added together in a single expression. A time series $\{x_t\}$ follows an autoregressive moving average (ARMA) process of order (p, q), denoted ARMA(p, q), when

$$x_t = \alpha_1 x_{t-1} + \alpha_2 x_{t-2} + \ldots + \alpha_p x_{t-p} + w_t + \beta_1 w_{t-1} + \beta_2 w_{t-2} + \ldots + \beta_q w_{t-q} \quad (6.7)$$

where $\{w_t\}$ is white noise. Equation (6.7) may be represented in terms of the backward shift operator $\mathbf{B}$ and rearranged in the more concise polynomial form

$$\theta_p(\mathbf{B})x_t = \phi_q(\mathbf{B})w_t \qquad (6.8)$$

The following points should be noted about an ARMA(p, q) process:

(a) The process is stationary when the roots of θ all exceed unity in absolute value.
(b) The process is invertible when the roots of ϕ all exceed unity in absolute value.
(c) The AR(p) model is the special case ARMA$(p, 0)$.
(d) The MA(q) model is the special case ARMA$(0, q)$.
(e) *Parameter parsimony.* When fitting to data, an ARMA model will often be more parameter efficient (i.e., require fewer parameters) than a single MA or AR model.

(e) *Parameter redundancy.* When θ and ϕ share a common factor, a stationary model can be simplified. For example, the model $(1 - \frac{1}{2}B)(1 - \frac{1}{3}B)x_t = (1 - \frac{1}{2}B)w_t$ can be written $(1 - \frac{1}{3}B)x_t = w_t$.

6.5.2 Derivation of second-order properties*

In order to derive the second-order properties for an ARMA(p, q) process $\{x_t\}$, it is helpful first to express the x_t in terms of white noise components w_t because white noise terms are independent. We illustrate the procedure for the ARMA(1, 1) model.

The ARMA(1, 1) process for $\{x_t\}$ is given by

$$x_t = \alpha x_{t-1} + w_t + \beta w_{t-1} \tag{6.9}$$

where w_t is white noise, with $E(w_t) = 0$ and $\mathrm{Var}(w_t) = \sigma_w^2$. Rearranging Equation (6.9) to express x_t in terms of white noise components,

$$x_t = (1 - \alpha B)^{-1}(1 + \beta B)w_t$$

Expanding the right-hand-side,

$$\begin{aligned}
x_t &= (1 + \alpha B + \alpha^2 B^2 + \ldots)(1 + \beta B)w_t \\
&= \left(\sum_{i=0}^{\infty} \alpha^i B^i\right)(1 + \beta B)\,w_t \\
&= \left(1 + \sum_{i=0}^{\infty} \alpha^{i+1} B^{i+1} + \sum_{i=0}^{\infty} \alpha^i \beta B^{i+1}\right)w_t \\
&= w_t + (\alpha + \beta)\sum_{i=1}^{\infty} \alpha^{i-1} w_{t-i} \tag{6.10}
\end{aligned}$$

With the equation in the form above, the second-order properties follow. For example, the mean $E(x_t)$ is clearly zero because $E(w_{t-i}) = 0$ for all i, and the variance is given by

$$\begin{aligned}
\mathrm{Var}(x_t) &= \mathrm{Var}\left[w_t + (\alpha + \beta)\sum_{i=1}^{\infty} \alpha^{i-1} w_{t-i}\right] \\
&= \sigma_w^2 + \sigma_w^2(\alpha + \beta)^2(1 - \alpha^2)^{-1} \tag{6.11}
\end{aligned}$$

The autocovariance γ_k, for $k > 0$, is given by

$$\begin{aligned}
\mathrm{Cov}\,(x_t, x_{t+k}) &= (\alpha + \beta)\,\alpha^{k-1}\sigma_w^2 + (\alpha + \beta)^2\,\sigma_w^2 \alpha^k \sum_{i=1}^{\infty} \alpha^{2i-2} \\
&= (\alpha + \beta)\,\alpha^{k-1}\sigma_w^2 + (\alpha + \beta)^2\,\sigma_w^2 \alpha^k (1 - \alpha^2)^{-1}
\end{aligned}$$

$$\tag{6.12}$$

The autocorrelation ρ_k then follows as

$$\rho_k = \gamma_k/\gamma_0 = \mathrm{Cov}\left(x_t, x_{t+k}\right)/\mathrm{Var}\left(x_t\right)$$
$$= \frac{\alpha^{k-1}(\alpha + \beta)(1 + \alpha\beta)}{1 + \alpha\beta + \beta^2} \tag{6.13}$$

Note that Equation (6.13) implies $\rho_k = \alpha\rho_{k-1}$.

6.6 ARMA models: Empirical analysis

6.6.1 Simulation and fitting

The ARMA process, and the more general ARIMA processes discussed in the next chapter, can be simulated using the R function `arima.sim`, which takes a list of coefficients representing the AR and MA parameters. An ARMA(p, q) model can be fitted using the `arima` function with the `order` function parameter set to `c(p, 0, q)`. The fitting algorithm proceeds similarly to that for an MA process. Below, data from an ARMA(1, 1) process are simulated for $\alpha = -0.6$ and $\beta = 0.5$ (Equation (6.7)), and an ARMA(1, 1) model fitted to the simulated series. As expected, the sample estimates of α and β are close to the underlying model parameters.

```
> set.seed(1)
> x <- arima.sim(n = 10000, list(ar = -0.6, ma = 0.5))
> coef(arima(x, order = c(1, 0, 1)))

      ar1       ma1  intercept
 -0.59697   0.50270  -0.00657
```

6.6.2 Exchange rate series

In §6.3, a simple MA(1) model failed to provide an adequate fit to the exchange rate series. In the code below, fitted MA(1), AR(1) and ARMA(1, 1) models are compared using the AIC. The ARMA(1, 1) model provides the best fit to the data, followed by the AR(1) model, with the MA(1) model providing the poorest fit. The correlogram in Figure 6.4 indicates that the residuals of the fitted ARMA(1, 1) model have small autocorrelations, which is consistent with a realisation of white noise and supports the use of the model.

```
> x.ma <- arima(x.ts, order = c(0, 0, 1))
> x.ar <- arima(x.ts, order = c(1, 0, 0))
> x.arma <- arima(x.ts, order = c(1, 0, 1))
> AIC(x.ma)

[1] -3.53

> AIC(x.ar)
```

```
[1] -37.4

> AIC(x.arma)

[1] -42.3

> x.arma

Call:
arima(x = x.ts, order = c(1, 0, 1))

Coefficients:
         ar1    ma1   intercept
       0.892  0.532      2.960
s.e.   0.076  0.202      0.244

sigma^2 estimated as 0.0151:  log likelihood = 25.1,   aic = -42.3

> acf(resid(x.arma))
```

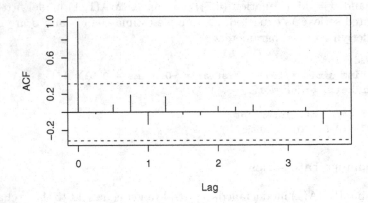

Fig. 6.4. The correlogram of residual series for the ARMA(1, 1) model fitted to the exchange rate data.

6.6.3 Electricity production series

Consider the Australian electricity production series introduced in §1.4.3. The data exhibit a clear positive trend and a regular seasonal cycle. Furthermore, the variance increases with time, which suggests a log-transformation may be appropriate (Fig. 1.5). A regression model is fitted to the logarithms of the original series in the code below.

```
> www <- "http://www.massey.ac.nz/~pscowper/ts/cbe.dat"
> CBE <- read.table(www, header = T)
> Elec.ts <- ts(CBE[, 3], start = 1958, freq = 12)
> Time <- 1:length(Elec.ts)
> Imth <- cycle(Elec.ts)
> Elec.lm <- lm(log(Elec.ts) ~ Time + I(Time^2) + factor(Imth))
> acf(resid(Elec.lm))
```

The correlogram of the residuals appears to cycle with a period of 12 months suggesting that the monthly indicator variables are not sufficient to account for the seasonality in the series (Fig. 6.5). In the next chapter, we find that this can be accounted for using a non-stationary model with a stochastic seasonal component. In the meantime, we note that the best fitting ARMA(p, q) model can be chosen using the smallest AIC either by trying a range of combinations of p and q in the arima function or using a for loop with upper bounds on p and q – taken as 2 in the code shown below. In each step of the for loop, the AIC of the fitted model is compared with the currently stored smallest value. If the model is found to be an improvement (i.e., has a smaller AIC value), then the new value and model are stored. To start with, best.aic is initialised to infinity (Inf). After the loop is complete, the best model can be found in best.order, and in this case the best model turns out to be an AR(2) model.

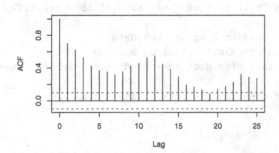

Fig. 6.5. Electricity production series: correlogram of the residual series of the fitted regression model.

```
> best.order <- c(0, 0, 0)
> best.aic <- Inf
> for (i in 0:2) for (j in 0:2) {
    fit.aic <- AIC(arima(resid(Elec.lm), order = c(i, 0,
        j)))
    if (fit.aic < best.aic) {
        best.order <- c(i, 0, j)
        best.arma <- arima(resid(Elec.lm), order = best.order)
        best.aic <- fit.aic
    }
```

```
  }
> best.order

[1] 2 0 0

> acf(resid(best.arma))
```

The `predict` function can be used both to forecast future values from the fitted regression model and forecast the future errors associated with the regression model using the ARMA model fitted to the residuals from the regression. These two forecasts can then be summed to give a forecasted value of the logarithm for electricity production, which would then need to be anti-logged and perhaps adjusted using a bias correction factor. As `predict` is a generic R function, it works in different ways for different input objects and classes. For a fitted regression model of class `lm`, the `predict` function requires the new set of data to be in the form of a data frame (object class `data.frame`). For a fitted ARMA model of class `arima`, the `predict` function requires just the number of time steps ahead for the desired forecast. In the latter case, `predict` produces an object that has both the predicted values and their standard errors, which can be extracted using `pred` and `se`, respectively. In the code below, the electricity production for each month of the next three years is predicted.

```
> new.time <- seq(length(Elec.ts), length = 36)
> new.data <- data.frame(Time = new.time, Imth = rep(1:12,
    3))
> predict.lm <- predict(Elec.lm, new.data)
> predict.arma <- predict(best.arma, n.ahead = 36)
> elec.pred <- ts(exp(predict.lm + predict.arma$pred), start = 1991,
    freq = 12)
> ts.plot(cbind(Elec.ts, elec.pred), lty = 1:2)
```

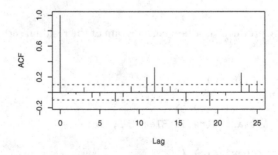

Fig. 6.6. Electricity production series: correlogram of the residual series of the best-fitting ARMA model.

The plot of the forecasted values suggests that the predicted values for winter may be underestimated by the fitted model (Fig. 6.7), which may be due to the remaining seasonal autocorrelation in the residuals (see Fig. 6.6). This problem will be addressed in the next chapter.

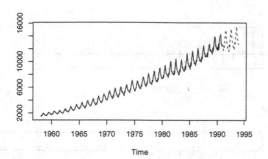

Fig. 6.7. Electricity production series: observed (solid line) and forecasted values (dotted line). The forecasted values are not likely to be accurate because of the seasonal autocorrelation present in the residuals for the fitted model.

6.6.4 Wave tank data

The data in the file `wave.dat` are the surface height of water (mm), relative to the still water level, measured using a capacitance probe positioned at the centre of a wave tank. The continuous voltage signal from this capacitance probe was sampled every 0.1 second over a 39.6-second period. The objective is to fit a suitable $ARMA(p, q)$ model that can be used to generate a realistic wave input to a mathematical model for an ocean-going tugboat in a computer simulation. The results of the computer simulation will be compared with tests using a physical model of the tugboat in the wave tank.

The pacf suggests that p should be at least 2 (Fig. 6.8). The best-fitting $ARMA(p, q)$ model, based on a minimum variance of residuals, was obtained with both p and q equal to 4. The acf and pacf of the residuals from this model are consistent with the residuals being a realisation of white noise (Fig. 6.9).

```
> www <- "http://www.massey.ac.nz/~pscowper/ts/wave.dat"
> wave.dat <- read.table(www, header = T)
> attach (wave.dat)
> layout(1:3)
> plot (as.ts(waveht), ylab = 'Wave height')
> acf (waveht)
> pacf (waveht)
> wave.arma <- arima(waveht, order = c(4,0,4))
> acf (wave.arma$res[-(1:4)])
> pacf (wave.arma$res[-(1:4)])
> hist(wave.arma$res[-(1:4)], xlab='height / mm', main='')
```

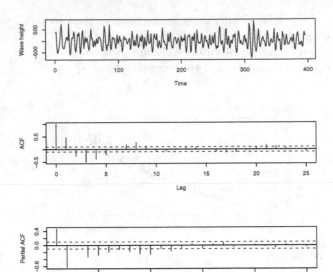

Fig. 6.8. Wave heights: time plot, acf, and pacf.

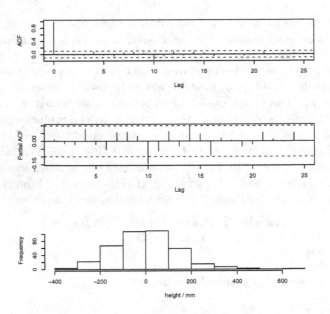

Fig. 6.9. Residuals after fitting an ARMA(4, 4) model to wave heights: acf, pacf, and histogram.

6.7 Summary of R commands

arima.sim	simulates data from an ARMA (or ARIMA) process
arima	fits an ARMA (or ARIMA) model to data
seq	generates a sequence
expression	used to plot maths symbol

6.8 Exercises

1. Using the relation $\text{Cov}(\sum x_t, \sum y_t) = \sum \sum \text{Cov}(x_t, y_t)$ (Equation (2.15)) for time series $\{x_t\}$ and $\{y_t\}$, prove Equation (6.3).

2. The series $\{w_t\}$ is white noise with zero mean and variance σ_w^2. For the following moving average models, find the autocorrelation function and determine whether they are invertible. In addition, simulate 100 observations for each model in R, compare the time plots of the simulated series, and comment on how the two series might be distinguished.
   a) $x_t = w_t + \frac{1}{2}w_{t-1}$
   b) $x_t = w_t + 2w_{t-1}$

3. Write the following models in ARMA(p, q) notation and determine whether they are stationary and/or invertible (w_t is white noise). In each case, check for parameter redundancy and ensure that the ARMA(p, q) notation is expressed in the simplest form.
   a) $x_t = x_{t-1} - \frac{1}{4}x_{t-2} + w_t + \frac{1}{2}w_{t-1}$
   b) $x_t = 2x_{t-1} - x_{t-2} + w_t$
   c) $x_t = \frac{3}{2}x_{t-1} - \frac{1}{2}x_{t-2} + w_t - \frac{1}{2}w_{t-1} + \frac{1}{4}w_{t-2}$
   d) $x_t = \frac{3}{2}x_{t-1} - \frac{1}{2}x_{t-2} + \frac{1}{2}w_t - w_{t-1}$
   e) $x_t = \frac{7}{10}x_{t-1} - \frac{1}{10}x_{t-2} + w_t - \frac{3}{2}w_{t-1}$
   f) $x_t = \frac{3}{2}x_{t-1} - \frac{1}{2}x_{t-2} + w_t - \frac{1}{3}w_{t-1} + \frac{1}{6}w_{t-2}$

4. a) Fit a suitable regression model to the air passenger series. Comment on the correlogram of the residuals from the fitted regression model.
   b) Fit an ARMA(p, q) model for values of p and q no greater than 2 to the residual series of the fitted regression model. Choose the best fitting model based on the AIC and comment on its correlogram.
   c) Forecast the number of passengers travelling on the airline in 1961.

5. a) Write an R function that calculates the autocorrelation function (Equation (6.13)) for an ARMA(1, 1) process. Your function should take parameters representing α and β for the AR and MA components.

b) Plot the autocorrelation function above for the case with $\alpha = 0.7$ and $\beta = -0.5$ for lags 0 to 20.

c) Simulate $n = 100$ values of the ARMA(1, 1) model with $\alpha = 0.7$ and $\beta = -0.5$, and compare the sample correlogram to the theoretical correlogram plotted in part (b). Repeat for $n = 1000$.

6. Let $\{x_t : t = 1, \ldots, n\}$ be a stationary time series with $E(x_t) = \mu$, $\text{Var}(x_t) = \sigma^2$, and $\text{Cor}(x_t, x_{t+k}) = \rho_k$. Using Equation (5.5) from Chapter 5:

a) Calculate $\text{Var}(\bar{x})$ when $\{x_t\}$ is the MA(1) process $x_t = w_t + \frac{1}{2}w_{t-1}$.

b) Calculate $\text{Var}(\bar{x})$ when $\{x_t\}$ is the MA(1) process $x_t = w_t - \frac{1}{2}w_{t-1}$.

c) Compare each of the above with the variance of the sample mean obtained for the white noise case $\rho_k = 0$ $(k > 0)$. Of the three models, which would have the most accurate estimate of μ based on the variances of their sample means?

d) A simulated example that extracts the variance of the sample mean for 100 Gaussian white noise series each of length 20 is given by
```
> set.seed(1)
> m <- rep(0, 100)
> for (i in 1:100) m[i] <- mean(rnorm(20))
> var(m)
[1] 0.0539
```
For each of the two MA(1) processes, write R code that extracts the variance of the sample mean of 100 realisations of length 20. Compare them with the variances calculated in parts (a) and (b).

7. If the sample autocorrelation function of a time series appears to cut off after lag q (i.e., autocorrelations at lags higher than q are not significantly different from 0 and do not follow any clear patterns), then an MA(q) model might be suitable. An AR(p) model is indicated when the *partial* autocorrelation function cuts off after lag p. If there are no convincing cutoff points for either function, an ARMA model may provide the best fit. Plot the autocorrelation and partial autocorrelation functions for the simulated ARMA(1, 1) series given in §6.6.1. Using the AIC, choose a best-fitting AR model and a best-fitting MA model. Which best-fitting model (AR or MA) has the smallest number of parameters? Compare this model with the fitted ARMA(1, 1) model of §6.6.1, and comment.

7

Non-stationary Models

7.1 Purpose

As we have discovered in the previous chapters, many time series are non-stationary because of seasonal effects or trends. In particular, random walks, which characterise many types of series, are non-stationary but can be transformed to a stationary series by first-order differencing (§4.4). In this chapter we first extend the random walk model to include autoregressive and moving average terms. As the differenced series needs to be aggregated (or 'integrated') to recover the original series, the underlying stochastic process is called autoregressive *integrated* moving average, which is abbreviated to ARIMA.

The ARIMA process can be extended to include seasonal terms, giving a non-stationary seasonal ARIMA (SARIMA) process. Seasonal ARIMA models are powerful tools in the analysis of time series as they are capable of modelling a very wide range of series. Much of the methodology was pioneered by Box and Jenkins in the 1970's.

Series may also be non-stationary because the variance is serially correlated (technically known as *conditionally heteroskedastic*), which usually results in periods of *volatility*, where there is a clear change in variance. This is common in financial series, but may also occur in other series such as climate records. One approach to modelling series of this nature is to use an autoregressive model for the variance, i.e. an autoregressive conditional heteroskedastic (ARCH) model. We consider this approach, along with the generalised ARCH (GARCH) model in the later part of the chapter.

7.2 Non-seasonal ARIMA models

7.2.1 Differencing and the electricity series

Differencing a series $\{x_t\}$ can remove trends, whether these trends are stochastic, as in a random walk, or deterministic, as in the case of a linear trend. In

P.S.P. Cowpertwait and A.V. Metcalfe, *Introductory Time Series with R*,   137
Use R, DOI 10.1007/978-0-387-88698-5_7,
© Springer Science+Business Media, LLC 2009

the case of a random walk, $x_t = x_{t-1} + w_t$, the first-order differenced series is white noise $\{w_t\}$ (i.e., $\nabla x_t = x_t - x_{t-1} = w_t$) and so is stationary. In contrast, if $x_t = a + bt + w_t$, a linear trend with white noise errors, then $\nabla x_t = x_t - x_{t-1} = b + w_t - w_{t-1}$, which is a stationary moving average process rather than white noise. Notice that the consequence of differencing a linear trend with white noise is an MA(1) process, whereas subtraction of the trend, $a + bt$, would give white noise. This raises an issue of whether or not it is sensible to use differencing to remove a deterministic trend. The arima function in R does not allow the fitted differenced models to include a constant. If you wish to fit a differenced model to a deterministic trend using R you need to difference, then mean adjust the differenced series to have a mean of 0, and then fit an ARMA model to the adjusted differenced series using arima with include.mean set to FALSE and d = 0.

A corresponding issue arises with simulations from an ARIMA model. Suppose $x_t = a + bt + w_t$ so $\nabla x_t = y_t = b + w_t - w_{t-1}$. It follows directly from the definitions that the inverse of $y_t = \nabla x_t$ is $x_t = x_0 + \sum_{i=1}^{t} y_i$. If an MA(1) model is fitted to the differenced time series, $\{y_t\}$, the coefficient of w_{t-1} is unlikely to be identified as precisely -1. It follows that the simulated $\{x_t\}$ will have increasing variance (Exercise 3) about a straight line.

We can take first-order differences in R using the difference function diff. For example, with the Australian electricity production series, the code below plots the data and first-order differences of the natural logarithm of the series. Note that in the layout command below the first figure is allocated two 1s and is therefore plotted over half (i.e., the first two fourths) of the frame.

```
> www <- "http://www.massey.ac.nz/~pscowper/ts/cbe.dat"
> CBE <- read.table(www, he = T)
> Elec.ts <- ts(CBE[, 3], start = 1958, freq = 12)
> layout(c(1, 1, 2, 3))
> plot(Elec.ts)
> plot(diff(Elec.ts))
> plot(diff(log(Elec.ts)))
```

The increasing trend is no longer apparent in the plots of the differenced series (Fig. 7.1).

7.2.2 Integrated model

A series $\{x_t\}$ is *integrated* of order d, denoted as I(d), if the dth difference of $\{x_t\}$ is white noise $\{w_t\}$; i.e., $\nabla^d x_t = w_t$. Since $\nabla^d \equiv (1 - \mathbf{B})^d$, where $\mathbf{B}$ is the backward shift operator, a series $\{x_t\}$ is integrated of order d if

$$(1 - \mathbf{B})^d x_t = w_t \tag{7.1}$$

The random walk is the special case I(1). The diff command from the previous section can be used to obtain higher-order differencing either by repeated application or setting the parameter d to the required values; e.g.,

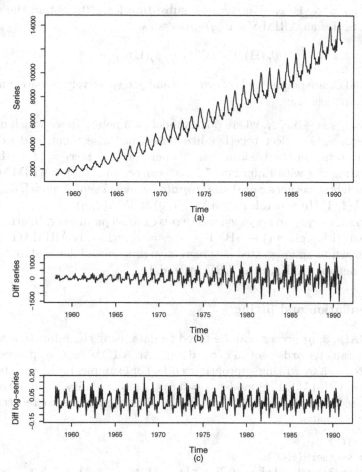

Fig. 7.1. (a) Plot of Australian electricity production series; (b) plot of the first-order differenced series; (c) plot of the first-order differenced log-transformed series.

`diff(diff(x))` and `diff(x, d=2)` would both produce second-order differenced series of `x`. Second-order differencing may sometimes successfully reduce a series with an underlying curve trend to white noise. A further parameter (`lag`) can be used to set the lag of the differencing. By default, `lag` is set to unity, but other values can be useful for removing additive seasonal effects. For example, `diff(x, lag=12)` will remove both a linear trend and additive seasonal effects in a monthly series.

7.2.3 Definition and examples

A time series $\{x_t\}$ follows an ARIMA(p, d, q) process if the dth differences of the $\{x_t\}$ series are an ARMA(p, q) process. If we introduce $y_t = (1 - \mathbf{B})^d x_t$, then $\theta_p(\mathbf{B})y_t = \phi_q(\mathbf{B})w_t$. We can now substitute for y_t to obtain the more succinct form for an ARIMA(p, d, q) process as

$$\theta_p(\mathbf{B})(1 - \mathbf{B})^d x_t = \phi_q(\mathbf{B})w_t \qquad (7.2)$$

where θ_p and ϕ_q are polynomials of orders p and q, respectively. Some examples of ARIMA models are:

(a) $x_t = x_{t-1}+w_t+\beta w_{t-1}$, where β is a model parameter. To see which model this represents, collect together like terms, factorise them, and express them in terms of the backward shift operator $(1 - \mathbf{B})x_t = (1 + \beta \mathbf{B})w_t$. Comparing this with Equation (7.2), we can see that $\{x_t\}$ is ARIMA(0, 1, 1), which is sometimes called an *integrated moving average* model, denoted as IMA(1, 1). In general, ARIMA(0, d, q) $\equiv$ IMA(d, q).

(b) $x_t = \alpha x_{t-1}+x_{t-1}-\alpha x_{t-2}+w_t$, where α is a model parameter. Rearranging and factorising gives $(1 - \alpha \mathbf{B})(1 - \mathbf{B})x_t = w_t$, which is ARIMA(1, 1, 0), also known as an integrated autoregressive process and denoted as ARI(1, 1). In general, ARI(p, d) $\equiv$ ARIMA(p, d, 0).

7.2.4 Simulation and fitting

An ARIMA(p, d, q) process can be fitted to data using the R function `arima` with the parameter `order` set to `c(p, d, q)`. An ARIMA(p, d, q) process can be simulated in R by writing appropriate code. For example, in the code below, data for the ARIMA(1, 1, 1) model $x_t = 0.5x_{t-1}+x_{t-1}-0.5x_{t-2}+w_t+0.3w_{t-1}$ are simulated and the model fitted to the simulated series to recover the parameter estimates.

```
> set.seed(1)
> x <- w <- rnorm(1000)
> for (i in 3:1000) x[i] <- 0.5 * x[i - 1] + x[i - 1] - 0.5 *
    x[i - 2] + w[i] + 0.3 * w[i - 1]
> arima(x, order = c(1, 1, 1))

Call:
arima(x = x, order = c(1, 1, 1))

Coefficients:
         ar1    ma1
       0.423  0.331
s.e.   0.043  0.045

sigma^2 estimated as 1.07:  log likelihood = -1450,  aic = 2906
```

Writing your own code has the advantage in that it helps to ensure that you understand the model. However, an ARIMA simulation can be carried out using the inbuilt R function `arima.sim`, which has the parameters `model` and `n` to specify the model and the simulation length, respectively.

```
> x <- arima.sim(model = list(order = c(1, 1, 1), ar = 0.5,
      ma = 0.3), n = 1000)
> arima(x, order = c(1, 1, 1))

Call:
arima(x = x, order = c(1, 1, 1))


Coefficients:
        ar1    ma1
      0.557  0.250
s.e.  0.037  0.044


sigma^2 estimated as 1.08:  log likelihood = -1457,  aic = 2921
```

7.2.5 IMA(1, 1) model fitted to the beer production series

The Australian beer production series is in the second column of the dataframe CBE in §7.2.1. The beer data is dominated by a trend of increasing beer production over the period, so a simple integrated model IMA(1, 1) is fitted to allow for this trend and a carryover of production from the previous month. The IMA(1, 1) model is often appropriate because it represents a linear trend with white noise added. The residuals are analysed using the correlogram (Fig. 7.2), which has peaks at yearly cycles and suggests that a seasonal term is required.

```
> Beer.ts <- ts(CBE[, 2], start = 1958, freq = 12)
> Beer.ima <- arima(Beer.ts, order = c(0, 1, 1))
> Beer.ima

Call:
arima(x = Beer.ts, order = c(0, 1, 1))


Coefficients:
         ma1
      -0.333
s.e.   0.056


sigma^2 estimated as 360:  log likelihood = -1723,  aic = 3451

> acf(resid(Beer.ima))
```

From the output above the fitted model is $x_t = x_{t-1} + w_t - 0.33w_{t-1}$. Forecasts can be obtained using this model, with t set to the value required for the forecast. Forecasts can also be obtained using the predict function in R with the parameter n.ahead set to the number of values in the future. For example, the production for the next year in the record is obtained using predict and the total annual production for 1991 obtained by summation:

```
> Beer.1991 <- predict(Beer.ima, n.ahead = 12)
> sum(Beer.1991$pred)
[1] 2365
```

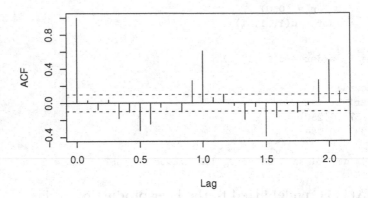

Fig. 7.2. Australian beer series: correlogram of the residuals of the fitted IMA(1, 1) model

7.3 Seasonal ARIMA models

7.3.1 Definition

A seasonal ARIMA model uses differencing at a lag equal to the number of seasons (s) to remove additive seasonal effects. As with lag 1 differencing to remove a trend, the lag s differencing introduces a moving average term. The seasonal ARIMA model includes autoregressive and moving average terms at lag s. The seasonal ARIMA$(p, d, q)(P, D, Q)_s$ model can be most succinctly expressed using the backward shift operator

$$\Theta_P(\mathbf{B}^s)\theta_p(\mathbf{B})(1 - \mathbf{B}^s)^D(1 - \mathbf{B})^d x_t = \Phi_Q(\mathbf{B}^s)\phi_q(\mathbf{B})w_t \qquad (7.3)$$

where Θ_P, θ_p, Φ_Q, and ϕ_q are polynomials of orders P, p, Q, and q, respectively. In general, the model is non-stationary, although if $D = d = 0$ and the roots of the characteristic equation (polynomial terms on the left-hand side of Equation (7.3)) all exceed unity in absolute value, the resulting model would be stationary. Some examples of seasonal ARIMA models are:

(a) A simple AR model with a seasonal period of 12 units, denoted as ARIMA$(0, 0, 0)(1, 0, 0)_{12}$, is $x_t = \alpha x_{t-12} + w_t$. Such a model would be appropriate for monthly data when only the value in the month of the previous year influences the current monthly value. The model is stationary when $|\alpha^{-1/12}| > 1$.

(b) It is common to find series with stochastic trends that nevertheless have seasonal influences. The model in (a) above could be extended to $x_t = x_{t-1} + \alpha x_{t-12} - \alpha x_{t-13} + w_t$. Rearranging and factorising gives

$(1 - \alpha \mathbf{B}^{12})(1 - \mathbf{B})x_t = w_t$ or $\Theta_1(\mathbf{B}^{12})(1 - \mathbf{B})x_t = w_t$, which, on comparing with Equation (7.3), is ARIMA$(0, 1, 0)(1, 0, 0)_{12}$. Note that this model could also be written $\nabla x_t = \alpha \nabla x_{t-12} + w_t$, which emphasises that the *change* at time t depends on the change at the same time (i.e., month) of the previous year. The model is non-stationary since the polynomial on the left-hand side contains the term $(1 - \mathbf{B})$, which implies that there exists a unit root $B = 1$.

(c) A simple quarterly seasonal moving average model is $x_t = (1 - \beta \mathbf{B}^4)w_t = w_t - \beta w_{t-4}$. This is stationary and only suitable for data without a trend. If the data also contain a stochastic trend, the model could be extended to include first-order differences, $x_t = x_{t-1} + w_t - \beta w_{t-4}$, which is an ARIMA$(0, 1, 0)(0, 0, 1)_4$ process. Alternatively, if the seasonal terms contain a stochastic trend, differencing can be applied at the seasonal period to give $x_t = x_{t-4} + w_t - \beta w_{t-4}$, which is ARIMA$(0, 0, 0)(0, 1, 1)_4$.

You should be aware that differencing at lag s will remove a linear trend, so there is a choice whether or not to include lag 1 differencing. If lag 1 differencing is included, when a linear trend is appropriate, it will introduce moving average terms into a white noise series. As an example, consider a time series of period 4 that is the sum of a linear trend, four additive seasonals, and white noise:

$$x_t = a + bt + s_{[t]} + w_t$$

where $[t]$ is the remainder after division of t by 4, so $s_{[t]} = s_{[t-4]}$. First, consider first-order differencing at lag 4 only. Then,

$$
\begin{aligned}
(1 - \mathbf{B}^4)x_t &= x_t - x_{t-4} \\
&= a + bt - (a + b(t - 4)) + s_{[t]} - s_{[t-4]} + w_t - w_{t-4} \\
&= 4b + w_t - w_{t-4}
\end{aligned}
$$

Formally, the model can be expressed as ARIMA$(0, 0, 0)(0, 1, 1)_4$ with a constant term $4b$. Now suppose we apply first-order differencing at lag 1 before differencing at lag 4. Then,

$$
\begin{aligned}
(1 - \mathbf{B}^4)(1 - \mathbf{B})x_t &= (1 - \mathbf{B}^4)(b + s_{[t]} - s_{[t-1]} + w_t - w_{t-1}) \\
&= w_t - w_{t-1} - w_{t-4} + w_{t-5}
\end{aligned}
$$

which is a ARIMA$(0, 1, 1)(0, 1, 1)_4$ model with no constant term.

7.3.2 Fitting procedure

Seasonal ARIMA models can potentially have a large number of parameters and combinations of terms. Therefore, it is appropriate to try out a wide range of models when fitting to data and to choose a best-fitting model using an appropriate criterion such as the AIC. Once a best-fitting model has been found, the correlogram of the residuals should be verified as white noise. Some confidence in the best-fitting model can be gained by deliberately overfitting the model by including further parameters and observing an increase in the AIC.

In R, this approach to fitting a range of seasonal ARIMA models is straight-forward, since the fitting criteria can be called by nesting functions and the 'up arrow' on the keyboard used to recall the last command, which can then be edited to try a new model. Any obvious terms, such as a differencing term if there is a trend, should be included and retained in the model to reduce the number of comparisons. The model can be fitted with the `arima` function, which requires an additional parameter `seasonal` to specify the seasonal com-ponents. In the example below, we fit two models with first-order terms to the logarithm of the electricity production series. The first uses autoregressive terms and the second uses moving average terms. The parameter $d = 1$ is re-tained in both the models since we found in §7.2.1 that first-order differencing successfully removed the trend in the series. The seasonal ARI model provides the better fit since it has the smallest AIC.

```
> AIC (arima(log(Elec.ts), order = c(1,1,0),
                  seas = list(order = c(1,0,0), 12)))
[1] -1765
> AIC (arima(log(Elec.ts), order = c(0,1,1),
                  seas = list(order = c(0,0,1), 12)))
[1] -1362
```

It is straightforward to check a range of models by a trial-and-error approach involving just editing a command on each trial to see if an improvement in the AIC occurs. Alternatively, we could write a simple function that fits a range of ARIMA models and selects the best-fitting model. This approach works better when the conditional sum of squares method CSS is selected in the `arima` function, as the algorithm is more robust. To avoid over parametrisation, the *consistent* Akaike Information Criteria (CAIC; see Bozdogan, 1987) can be used in model selection. An example program follows.

```
get.best.arima <- function(x.ts, maxord = c(1,1,1,1,1,1))
{
  best.aic <- 1e8
  n <- length(x.ts)
  for (p in 0:maxord[1]) for(d in 0:maxord[2]) for(q in 0:maxord[3])
    for (P in 0:maxord[4]) for(D in 0:maxord[5]) for(Q in 0:maxord[6])
    {
      fit <- arima(x.ts, order = c(p,d,q),
                          seas = list(order = c(P,D,Q),
                          frequency(x.ts)), method = "CSS")
      fit.aic <- -2 * fit$loglik + (log(n) + 1) * length(fit$coef)
      if (fit.aic < best.aic)
      {
```

```
        best.aic <- fit.aic
        best.fit <- fit
        best.model <- c(p,d,q,P,D,Q)
     }
  }
  list(best.aic, best.fit, best.model)
}

> best.arima.elec <- get.best.arima( log(Elec.ts),
                              maxord = c(2,2,2,2,2,2))
> best.fit.elec <- best.arima.elec[[2]]
> acf( resid(best.fit.elec) )
> best.arima.elec [[3]]

[1] 0 1 1 2 0 2

> ts.plot( cbind( window(Elec.ts,start = 1981),
                  exp(predict(best.fit.elec,12)$pred) ), lty = 1:2)
```

From the code above, we see the best-fitting model using terms up to second order is $ARIMA(0, 1, 1)(2, 0, 2)_{12}$. Although higher-order terms could be tried by increasing the values in `maxord`, this would seem unnecessary since the residuals are approximately white noise (Fig. 7.3b). For the predicted values (Fig. 7.3a), a biased correction factor could be used, although this would seem unnecessary given that the residual standard deviation is small compared with the predictions.

7.4 ARCH models

7.4.1 S&P500 series

Standard and Poors (of the McGraw-Hill companies) publishes a range of financial indices and credit ratings. Consider the following time plot and correlogram of the daily returns of the S&P500 Index[1] (from January 2, 1990 to December 31, 1999), available in the MASS library within R.

```
> library(MASS)
> data(SP500)
> plot(SP500, type = 'l')
> acf(SP500)
```

The time plot of the returns is shown in Figure 7.4(a), and at first glance the series appears to be a realisation of a stationary process. However, on

---

[1] The S&P500 Index is calculated from the stock prices of 500 large corporations. The time series in R is the returns of the S&P500 Index, defined as $100\ln(SPI_t/SPI_{t-1})$, where SPI_t is the value of the S&P500 Index on trading day t.

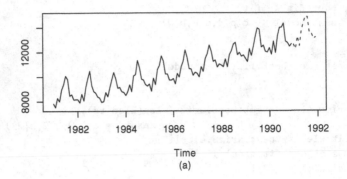

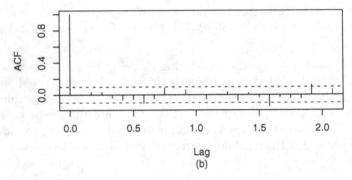

Fig. 7.3. Electricity production series: (a) time plot for last 10 years, with added predicted values (dotted); (b) correlogram of the residuals of the best-fitting seasonal ARIMA model.

closer inspection, it seems that the variance is smallest in the middle third of the series and greatest in the last third. The series exhibits periods of increased variability, sometimes called *volatility* in the financial literature, although it does not increase in a regular way. When a variance is not constant in time but changes in a regular way, as in the airline and electricity data (where the variance increased with the trend), the series is called *heteroskedastic*. If a series exhibits periods of increased variance, so the variance is correlated in time (as observed in the S&P500 data), the series exhibits volatility and is called *conditional heteroskedastic*.

Note that the correlogram of a volatile series does not differ significantly from white noise (Fig. 7.4b), but the series is non-stationary since the variance is different at different times. If a correlogram appears to be white noise (e.g., Fig. 7.4b), then volatility can be detected by looking at the correlogram of the squared values since the squared values are equivalent to the variance

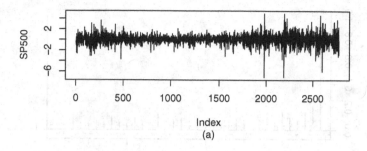

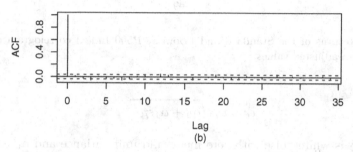

Fig. 7.4. Standard and Poors returns of the S&P500 Index: (a) time plot; (b) correlogram.

(provided the series is adjusted to have a mean of zero). The mean of the returns of the S&P500 Index between January 2, 1990 and December 31, 1999 is 0.0458. Although this is small compared with the variance, it accounts for an increase in the S&P500 Index from 360 to 1469 over the 2527 trading days. The correlogram of the squared mean-adjusted values of the S&P500 index is given by

```
> acf((SP500 - mean(SP500))^2)
```

From this we can see that there is evidence of serial correlation in the squared values, so there is evidence of conditional heteroskedastic behaviour and volatility (Fig. 7.5).

7.4.2 Modelling volatility: Definition of the ARCH model

In order to account for volatility, we require a model that allows for conditional changes in the variance. One approach to this is to use an autoregressive model for the variance process. This leads to the following definition. A series $\{\epsilon_t\}$ is first-order autoregressive conditional heteroskedastic, denoted ARCH(1), if

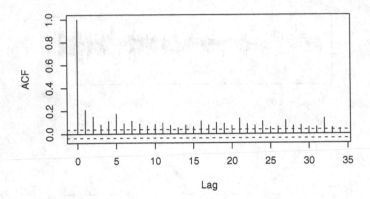

Fig. 7.5. Returns of the Standard and Poors S&P500 Index: correlogram of the squared mean-adjusted values.

$$\epsilon_t = w_t \sqrt{\alpha_0 + \alpha_1 \epsilon_{t-1}^2} \qquad (7.4)$$

where $\{w_t\}$ is white noise with zero mean and unit variance and α_0 and α_1 are model parameters.

To see how this introduces volatility, square Equation (7.4) to calculate the variance

$$
\begin{aligned}
\mathrm{Var}\,(\epsilon_t) &= E\left(\epsilon_t^2\right) \\
&= E\left(w_t^2\right) E\left(\alpha_0 + \alpha_1 \epsilon_{t-1}^2\right) \\
&= E\left(\alpha_0 + \alpha_1 \epsilon_{t-1}^2\right) \\
&= \alpha_0 + \alpha_1 \mathrm{Var}\,(\epsilon_{t-1}) \qquad (7.5)
\end{aligned}
$$

since $\{w_t\}$ has unit variance and $\{\epsilon_t\}$ has zero mean. If we compare Equation (7.5) with the AR(1) process $x_t = \alpha_0 + \alpha_1 x_{t-1} + w_t$, we see that the variance of an ARCH(1) process behaves just like an AR(1) model. Hence, in model fitting, a decay in the autocorrelations of the *squared* residuals should indicate whether an ARCH model is appropriate or not. The model should only be applied to a prewhitened residual series $\{\epsilon_t\}$ that is uncorrelated and contains no trends or seasonal changes, such as might be obtained after fitting a satisfactory SARIMA model.

7.4.3 Extensions and GARCH models

The first-order ARCH model can be extended to a pth-order process by including higher lags. An ARCH(p) process is given by

$$\epsilon_t = w_t \sqrt{\alpha_0 + \sum_{i=1}^{p} \alpha_p \epsilon_{t-i}^2} \qquad (7.6)$$

where $\{w_t\}$ is again white noise with zero mean and unit variance.

A further extension, widely used in financial applications, is the generalised ARCH model, denoted GARCH(q, p), which has the ARCH(p) model as the special case GARCH(0, p). A series $\{\epsilon_t\}$ is GARCH(q, p) if

$$\epsilon_t = w_t \sqrt{h_t} \qquad (7.7)$$

where

$$h_t = \alpha_0 + \sum_{i=1}^{p} \alpha_i \epsilon_{t-i}^2 + \sum_{j=1}^{q} \beta_j h_{t-j} \qquad (7.8)$$

and α_i and β_j ($i = 0, 1, \ldots, p;\ j = 1, \ldots, q$) are model parameters. In R, a GARCH model can be fitted using the `garch` function in the `tseries` library (Trapletti and Hornik, 2008). An example now follows.

7.4.4 Simulation and fitted GARCH model

In the following code data are simulated for the GARCH(1, 1) model $a_t = w_t\sqrt{h_t}$, where $h_t = \alpha_0 + \alpha_1 a_{t-1} + \beta_1 h_{t-1}$ with $\alpha_1 + \beta_1 < 1$ to ensure stability; e.g., see Enders (1995). The simulated series are placed in the vector **a** and the correlograms plotted (Fig. 7.6).

```
> set.seed(1)
> alpha0 <- 0.1
> alpha1 <- 0.4
> beta1 <- 0.2
> w <- rnorm(10000)
> a <- rep(0, 10000)
> h <- rep(0, 10000)
> for (i in 2:10000) {
    h[i] <- alpha0 + alpha1 * (a[i - 1]^2) + beta1 * h[i -
        1]
    a[i] <- w[i] * sqrt(h[i])
  }
> acf(a)
> acf(a^2)
```

The series in **a** exhibits the GARCH characteristics of uncorrelated values (Fig. 7.6a) but correlated squared values (Fig. 7.6b).

In the following example, a GARCH model is fitted to the simulated series using the `garch` function, which can be seen to recover the original parameters since these fall within the 95% confidence intervals. The default is GARCH(1, 1), which often provides an adequate model, but higher-order models can be specified with the parameter `order=c(p,q)` for some choice of p and q.

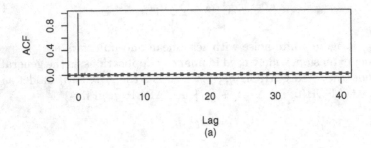

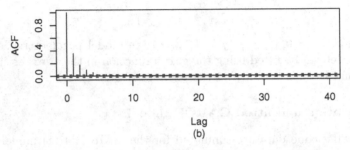

Fig. 7.6. Correlograms for GARCH series: (a) simulated series; (b) squared values of simulated series.

```
> library(tseries)

> a.garch <- garch(a, grad = "numerical", trace = FALSE)
> confint(a.garch)

    2.5 % 97.5 %
a0 0.0882  0.109
a1 0.3308  0.402
b1 0.1928  0.295
```

In the example above, we have used the parameter `trace=F` to suppress output and a numerical estimate of gradient `grad="numerical"` that is slightly more robust (in the sense of algorithmic convergence) than the default.

7.4.5 Fit to S&P500 series

The GARCH model is fitted to the S&P500 return series. The residual series of the GARCH model $\{\hat{w}_t\}$ are calculated from

$$\hat{w}_t = \frac{\epsilon_t}{\sqrt{\hat{h}_t}}$$

If the GARCH model is suitable the residual series should appear to be a realisation of white noise with zero mean and unit variance. In the case of a GARCH(1, 1) model,

$$\hat{h}_t = \hat{\alpha}_0 + \hat{\alpha}_1 \epsilon_{t-1}^2 + \hat{\beta}_1 \hat{h}_{t-1}$$

with $\hat{h}_1 = 0$ for $t = 2, \ldots, n$.[2] The calculations are performed by the function garch. The first value in the residual series is not available (NA), so we remove the first value using [-1] and the correlograms are then found for the resultant residual and squared residual series:

```
> sp.garch <- garch(SP500, trace = F)
> sp.res    <- sp.garch$res[-1]
> acf(sp.res)
> acf(sp.res^2)
```

Both correlograms suggest that the residuals of the fitted GARCH model behave like white noise, indicating a satisfactory fit has been obtained (Fig. 7.7).

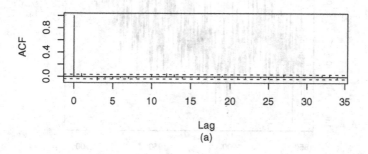

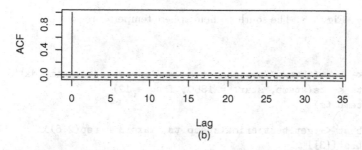

Fig. 7.7. GARCH model fitted to mean-adjusted S&P500 returns: (a) correlogram of the residuals; (b) correlogram of the squared residuals.

---

[2] Notice that a residual for time $t = 1$ cannot be calculated from this formula.

7.4.6 Volatility in climate series

Recently there have been studies on volatility in climate series (e.g., Romilly, 2005). Temperature data (1850–2007; see Brohan et al. 2006) for the southern hemisphere were extracted from the database maintained by the University of East Anglia Climatic Research Unit and edited into a form convenient for reading into R. In the following code, the series are read in, plotted (Fig. 7.8), and a best-fitting seasonal ARIMA model obtained using the get.best.arima function given in §7.3.2. Confidence intervals for the parameters were then evaluated (the transpose t() was taken to provide these in rows instead of columns).

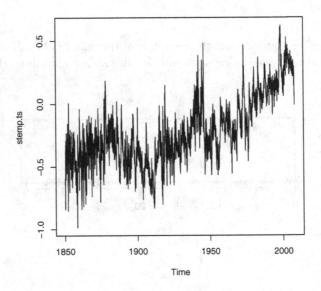

Fig. 7.8. The southern hemisphere temperature series.

```
> stemp <- scan("http://www.massey.ac.nz/~pscowper/ts/stemp.dat")
> stemp.ts <- ts(stemp, start = 1850, freq = 12)
> plot(stemp.ts)

> stemp.best <- get.best.arima(stemp.ts, maxord = rep(2,6))
> stemp.best[[3]]

[1] 1 1 2 2 0 1

> stemp.arima <- arima(stemp.ts, order = c(1,1,2),
                        seas = list(order = c(2,0,1), 12))
```

```
> t( confint(stemp.arima) )


          ar1    ma1    ma2  sar1     sar2  sma1
2.5 %   0.832  -1.45  0.326 0.858  -0.0250 -0.97
97.5 %  0.913  -1.31  0.453 1.004   0.0741 -0.85
```

The second seasonal AR component is not significantly different from zero, and therefore the model is refitted leaving this component out:

```
> stemp.arima <- arima(stemp.ts, order = c(1,1,2),
                             seas = list(order = c(1,0,1), 12))

> t( confint(stemp.arima) )

         ar1    ma1    ma2   sar1    sma1
2.5 %   0.83  -1.45  0.324  0.924  -0.969
97.5 %  0.91  -1.31  0.451  0.996  -0.868
```

To check for goodness-of-fit, the correlogram of residuals from the ARIMA model is plotted (Fig. 7.9a). In addition, to investigate volatility, the correlogram of the squared residuals is found (Fig. 7.9b).

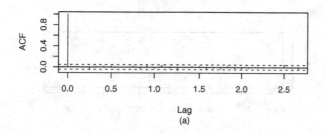

(a)

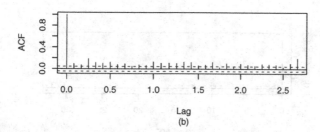

(b)

Fig. 7.9. Seasonal ARIMA model fitted to the temperature series: (a) correlogram of the residuals; (b) correlogram of the squared residuals.

```
> stemp.res <- resid(stemp.arima)
> layout(1:2)
```

```
> acf(stemp.res)
> acf(stemp.res^2)
```

There is clear evidence of volatility since the squared residuals are correlated at most lags (Fig. 7.9b). Hence, a GARCH model is fitted to the residual series:

```
> stemp.garch <- garch(stemp.res, trace = F)
> t(confint(stemp.garch))

           a0      a1     b1
2.5 %  1.06e-05 0.0330 0.925
97.5 % 1.49e-04 0.0653 0.963

> stemp.garch.res <- resid(stemp.garch)[-1]
> acf(stemp.garch.res)
> acf(stemp.garch.res^2)
```

Based on the output above, we can see that the coefficients of the fitted GARCH model are all statistically significant, since zero does not fall in any of the confidence intervals. Furthermore, the correlogram of the residuals shows no obvious patterns or significant values (Fig. 7.10). Hence, a satisfactory fit has been obtained.

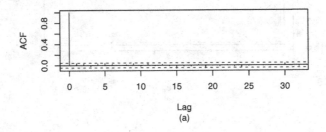

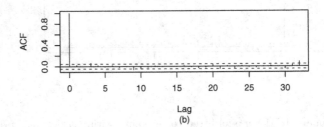

Fig. 7.10. GARCH model fitted to the residuals of the seasonal ARIMA model of the temperature series: (a) correlogram of the residuals; (b) correlogram of the squared residuals.

7.4.7 GARCH in forecasts and simulations

If a GARCH model is fitted to the residual errors of a fitted time series model, it will not influence the *average* prediction at some point in time since the mean of the residual errors is zero. Thus, single-point forecasts from a fitted time series model remain unchanged when GARCH models are fitted to the residuals. However, a fitted GARCH model will affect the variance of simulated predicted values and thus result in periods of changing variance or volatility in simulated series.

The main application of GARCH models is for simulation studies, especially in finance, insurance, teletraffic, and climatology. In all these applications, the periods of high variability tend to lead to untoward events, and it is essential to model them in a realistic manner. Typical R code for simulation was given in §7.4.4.

7.5 Summary of R commands

`garch`   fits a GARCH (or ARCH) model to data

7.6 Exercises

In each of the following, $\{w_t\}$ is white noise with zero mean.

1. Identify each of the following as specific ARIMA models and state whether or not they are stationary.

   a) $z_t = z_{t-1} - 0.25z_{t-2} + w_t + 0.5w_{t-1}$

   b) $z_t = 2z_{t-1} - z_{t-2} + w_t$

   c) $z_t = 0.5z_{t-1} + 0.5z_{t-2} + w_t - 0.5w_{t-1} + 0.25w_{t-2}$

2. Identify the following as certain multiplicative seasonal ARIMA models and find out whether they are invertible and stationary.

   a) $z_t = 0.5z_{t-1} + z_{t-4} - 0.5z_{t-5} + w_t - 0.3w_{t-1}$

   b) $z_t = z_{t-1} + z_{t-12} - z_{t-13} + w_t - 0.5w_{t-1} - 0.5w_{t-12} + 0.25w_{t-13}$

3. Suppose $x_t = a + bt + w_t$. Define $y_t = \nabla x_t$.

   a) Show that $x_t = x_0 + \sum_{i=1}^{t} y_i$ and identify x_0.

   b) Now suppose an MA(1) model is fitted to $\{y_t\}$ and the fitted model is $y_t = b + w_t + \beta w_{t-1}$. Show that a simulated $\{x_t\}$ will have increasing variance about the line $a + bt$ unless β is precisely -1.

4. The number of overseas visitors to New Zealand is recorded for each month over the period 1977 to 1995 in the file osvisit.dat on the book website (http://www.massey.ac.nz/~pscowper/ts/osvisit.dat). Download the file into R and carry out the following analysis. Your solution should include any R commands, plots, and comments. Let x_t be the number of overseas visitors in time period t (in months) and $z_t = \ln(x_t)$.

a) Comment on the main features in the correlogram for $\{z_t\}$.

b) Fit an ARIMA(1, 1, 0) model to $\{z_t\}$ giving the estimated AR parameter and the standard deviation of the residuals. Comment on the correlogram of the residuals of this fitted ARIMA model.

c) Fit a seasonal ARIMA(1, 1, 0)(0, 1, 0)$_{12}$ model to $\{z_t\}$ and plot the correlogram of the residuals of this model. Has seasonal differencing removed the seasonal effect? Comment.

d) Choose the best-fitting Seasonal ARIMA model from the following: ARIMA(1, 1, 0)(1, 1, 0)$_{12}$, ARIMA(0, 1, 1)(0, 1, 1)$_{12}$, ARIMA(1, 1, 0)(0, 1, 1)$_{12}$, ARIMA(0, 1, 1)(1, 1, 0)$_{12}$, ARIMA(1, 1, 1)(1, 1, 1)$_{12}$, ARIMA(1, 1, 1)(1, 1, 0)$_{12}$, ARIMA(1, 1, 1)(0, 1, 1)$_{12}$. Base your choice on the AIC, and comment on the correlogram of the residuals of the best-fitting model.

e) Express the best-fitting model in part (d) above in terms of z_t, white noise components, and the backward shift operator (you will need to write this out by hand, but it is not necessary to expand all the factors).

f) Test the residuals from the best-fitting seasonal ARIMA model for stationarity.

g) Forecast the number of overseas visitors for each month in the next year (1996), and give the total number of visitors expected in 1996 under the fitted model. [Hint: To get the forecasts, you will need to use the exponential function of the generated seasonal ARIMA forecasts and multiply these by a bias correction factor based on the mean square residual error.]

5. Use the get.best.arima function from §7.3.2 to obtain a best-fitting ARIMA$(p, d, q)(P, D, Q)_{12}$ for all $p, d, q, P, D, Q \leq 2$ to the logarithm of the Australian chocolate production series (in the file at http://www.massey.ac.nz/~pscowper/ts/cbe.dat). Check that the correlogram of the residuals for the best-fitting model is representative of white noise. Check the correlogram of the squared residuals. Comment on the results.

6. This question uses the data in stockmarket.dat on the book website http://www.massey.ac.nz/~pscowper/ts/, which contains stock market

data for seven cities for the period January 6, 1986 to December 31, 1997. Download the data into R and put the data into a variable x. The first three rows should be:

```
> x[1:3,]
  Amsterdam Frankfurt London HongKong  Japan Singapore NewYork
1    275.76   1425.56 1424.1  1796.59 13053.8    233.63  210.65
2    275.43   1428.54 1415.2  1815.53 12991.2    237.37  213.80
3    278.76   1474.24 1404.2  1826.84 13056.4    240.99  207.97
```

a) Plot the Amsterdam series and the first-order differences of the series. Comment on the plots.
b) Fit the following models to the Amsterdam series, and select the best fitting model: ARIMA(0, 1, 0); ARIMA(1, 1, 0), ARIMA(0, 1, 1), ARIMA(1, 1, 1).
c) Produce the correlogram of the residuals of the best-fitting model and the correlogram of the squared residuals. Comment.
d) Fit the following GARCH models to the residuals, and select the best-fitting model: GARCH(0, 1), GARCH(1, 0), GARCH(1, 1), and GARCH(0, 2). Give the estimated parameters of the best-fitting model.
e) Plot the correlogram of the residuals from the best fitting GARCH model. Plot the correlogram of the squared residuals from the best fitting GARCH model, and comment on the plot.

7. Predict the monthly temperatures for 2008 using the model fitted to the climate series in §7.4.6, and add these predicted values to a time plot of the temperature series from 1990. Give an upper bound for the predicted values based on a 95% confidence level. Simulate ten possible future temperature scenarios for 2008. This will involve generating GARCH errors and adding these to the predicted values from the fitted seasonal ARIMA model.

8

Long-Memory Processes

8.1 Purpose

Some time series exhibit marked correlations at high lags, and they are referred to as long-memory processes. Long-memory is a feature of many geophysical time series. Flows in the Nile River have correlations at high lags, and Hurst (1951) demonstrated that this affected the optimal design capacity of a dam. Mudelsee (2007) shows that long-memory is a hydrological property that can lead to prolonged drought or temporal clustering of extreme floods. At a rather different scale, Leland et al. (1993) found that Ethernet local area network (LAN) traffic appears to be statistically self-similar and a long-memory process. They showed that the nature of congestion produced by self-similar traffic differs drastically from that predicted by the traffic models used at that time. Mandelbrot and co-workers investigated the relationship between self-similarity and long term memory and played a leading role in establishing fractal geometry as a subject of study.

8.2 Fractional differencing

Beran (1994) describes the qualitative features of a typical sample path (realisation) from a long-memory process. There are relatively long periods during which the observations tend to stay at a high level and similar long periods during which observations tend to be at a low level. There may appear to be trends or cycles over short time periods, but these do not persist and the entire series looks stationary. A more objective criterion is that sample correlations r_k decay to zero at a rate that is approximately proportional to $k^{-\lambda}$ for some $0 < \lambda < 1$. This is noticeably slower than the rate of decay of r_k for realisations from an AR(1) process, for example, which is approximately proportional to λ^k for some $0 < \lambda < 1$.

The mathematical definition of a stationary process with long-memory, also known as long-range dependence or persistence, can be given in terms of

P.S.P. Cowpertwait and A.V. Metcalfe, *Introductory Time Series with R*,   159
Use R, DOI 10.1007/978-0-387-88698-5_8,
© Springer Science+Business Media, LLC 2009

its autocorrelation function. A stationary process x_t with long-memory has an autocorrelation function ρ_k that satisfies the condition

$$\lim_{k \to \infty} \rho_k = ck^{-\lambda}$$

for some $0 < c$ and $0 < \lambda < 1$. The closer λ is to 0, the more pronounced is the long-memory.

The hydrologist Harold Hurst found that for many geophysical records, including the Nile River data, a statistic known as the rescaled range (Exercise 4) over a period k is approximately proportional to k^H for some $H > \frac{1}{2}$. The Hurst parameter, H, is defined by $H = 1 - \lambda/2$ and hence ranges from $\frac{1}{2}$ to 1. The closer H is to 1, the more persistent the time series. If there is no long-memory effect, then $H = \frac{1}{2}$.

A fractionally differenced ARIMA process $\{x_t\}$, FARIMA(p, d, q), has the form

$$\phi(\mathbf{B})(1 - \mathbf{B})^d x_t = \psi(\mathbf{B})w_t \tag{8.1}$$

for some $-\frac{1}{2} < d < \frac{1}{2}$. The range $0 < d < \frac{1}{2}$ gives long-memory processes. It can be useful to introduce the fractionally differenced series $\{y_t\}$ and express Equation (8.1) as

$$y_t = (1 - \mathbf{B})^d x_t = [\phi(\mathbf{B})]^{-1}\psi(\mathbf{B})w_t \tag{8.2}$$

because this suggests a means of fitting a FARIMA model to time series. For a trial value of d, we calculate the fractionally differenced series $\{y_t\}$, fit an ARIMA model to $\{y_t\}$, and then investigate the residuals. The calculation of the fractionally differenced series $\{y_t\}$ follows from a formal binomial expansion of $(1 - B)^d$ and is given by

$$(1 - B)^d = 1 - dB + \frac{d(d - 1)}{2!}B^2 - \frac{d(d - 1)(d - 2)}{3!}B^3 + \cdots$$

curtailed at some suitably large lag (L), which might reasonably be set to 40. For example, if $d = 0.45$, then

$$y_t = x_t - 0.450x_{t-1} - 0.12375x_{t-2} - 0.0639375x_{t-3} - \cdots - 0.001287312x_{t-40}$$

The R code for calculating the coefficients is

```
> cf <- rep(0,40)
> d <- 0.45
> cf[1] <- -d
> for (i in 1:39) cf[i+1] <- -cf[i] * (d-i) / (i+1)
```

Another equivalent expression for Equation (8.1), which is useful for simulations, is

$$x_t = [\phi(\mathbf{B})]^{-1}\psi(\mathbf{B})(1 - \mathbf{B})^{-d}w_t$$

In simulations, the first step is to calculate $(1-B)^{-d}w_t$. The operator $(1-B)^{-d}$ needs to be expanded as

$$(1 - B)^{-d} = 1 - d(-B) + \frac{-d(-d-1)}{2!}B^2 - \frac{-d(-d-1)(-d-2)}{3!}B^3 + \cdots$$

with the series curtailed at some suitably large lag L. The distributions for the independent white noise series can be chosen to fit the application, and in finance and telecommunications, heavy-tailed distributions are often appropriate. In particular, a t-distribution with ν (>4) degrees of freedom has kurtosis $6/(\nu-4)$ and so is heavy tailed. If, for example, $d = 0.45$ and $L = 40$, then

$$(1 - B)^{-d}w_t = w_t + 0.45w_{t-1} + 0.32625w_{t-2} + 0.2664375w_{t-3}$$
$$+ \cdots + 0.0657056w_{t-40}$$

The autocorrelation function ρ_k of a FARIMA$(0, d, 0)$ process tends towards

$$\frac{\Gamma(1-d)}{\Gamma(d)}|k|^{2d-1}$$

for large n. The process is stationary provided $-\frac{1}{2} < d < \frac{1}{2}$. This provides a relationship between the differencing parameter d and the long-memory parameter λ when $0 \le d$:

$$2d - 1 = -\lambda \iff d = \frac{1-\lambda}{2}$$

A FARIMA$(0, d, 0)$ model, with $0 < d < \frac{1}{2}$, lies between a stationary AR(1) model and a non-stationary random walk. In practice, for fitting or simulation, we have to truncate a FARIMA$(0, d, 0)$ process at some lag L. Then it is equivalent to an AR(L) model, but all the coefficients in the FARIMA$(0, d, 0)$ model depend on the single parameter d.

8.3 Fitting to simulated data

In the following script, the function `fracdiff.sim` generates a realisation from a FARIMA process.[1] The first parameter is the length of the realisation, and then AR and MA parameters can be specified – use `c()` if there is more than one of each, followed by a value for d. The default for the discrete white noise (DWN) component is standard Gaussian, but this can be varied by using **innov** or **rand.gen**, as described in `help(fracdiff.sim)`. We then fit a FARIMA model to the realisation. In this case, we set the number of AR coefficients to be fitted to 1, but when fitting to a time series from an unknown model, we should try several values for the number of autoregressive and moving average parameters (**nar** and **nma**, respectively).

---

[1] You will need to have the `fracdiff` library installed. This can be downloaded from CRAN.

```
> library(fracdiff)
> set.seed(1)
> fds.sim <- fracdiff.sim(10000, ar = 0.9, d = 0.4)
> x <- fds.sim$series
> fds.fit <- fracdiff(x, nar = 1)
```

In the code below, the first for loop calculates the coefficients for the lagged terms in the fractional differences using the fitted value for d. The following nested loop then calculates the fractionally differenced time series. Then an AR model is fitted to the differenced series and the acf for the residuals is plotted (Fig. 8.2). The residuals should appear to be a realisation of DWN.

```
> n <- length(x)
> L <- 30
> d <- fds.fit$d
> fdc <- d
> fdc[1] <- fdc
> for (k in 2:L) fdc[k] <- fdc[k-1] * (d+1-k) / k
> y <- rep(0, L)
> for (i in (L+1):n) {
    csm <- x[i]
    for (j in 1:L) csm <- csm + ((-1)^j) * fdc[j] * x[i-j]
    y[i] <- csm
  }
> y <- y[(L+1):n]
> z.ar <- ar(y)
> ns <- 1 + z.ar$order
> z <- z.ar$res [ns:length(y)]
> par(mfcol = c(2, 2))
> plot(as.ts(x), ylab = "x")
> acf(x) ; acf(y) ; acf(z)
```

In Figure 8.1, we show the results when we generate a realisation $\{x_t\}$ from a fractional difference model with no AR or MA parameters, FARIMA(0, 0.4, 0). The very slow decay in both the acf and pacf indicates long-memory. The estimate of d is 0.3921. The fractionally differenced series, $\{y_t\}$, appears to be a realisation of DWN. If, instead of fitting a FARIMA(0, d, 0) model, we use ar, the order selected is 38. The residuals from AR(38) also appear to be a realisation from DWN, but the single-parameter FARIMA model is far more parsimonious.

In Figure 8.2, we show the results when we generate a realisation $\{x_t\}$ from a FARIMA(1, 0.4, 0) model with an AR parameter of 0.9. The estimates of d and the AR parameter, obtained from fracdiff, are 0.429 and 0.884, respectively. The estimate of the AR parameter made from the fractionally differenced series $\{y_t\}$ using ar is 0.887, and the slight difference is small by comparison with the estimated error and is of no practical importance. The residuals appear to be a realisation of DWN (Fig. 8.2).

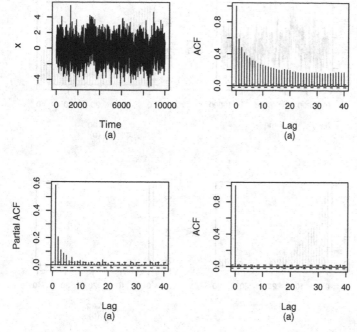

Fig. 8.1. A simulated series with long-memory FARIMA(0, 0.4, 0): (a) time series plot (**x**); (b) correlogram of series **x**; (c) partial correlogram of **y**; (d) correlogram after fractional differencing (**z**).

```
> summary(fds.fit)

...

Coefficients:
    Estimate Std. Error z value Pr(>|z|)
d    0.42904    0.01439    29.8   <2e-16 ***
ar   0.88368    0.00877   100.7   <2e-16 ***
ma   0.00000    0.01439     0.0        1
...

> ar(y)

Coefficients:
    1
0.887

Order selected 1  sigma^2 estimated as  1.03
```

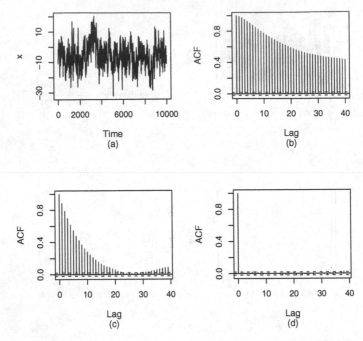

Fig. 8.2. A time series with long-memory FARIMA(1, 0.4, 0): (a) time series plot (**x**); (b) correlogram of series **x**; (c) correlogram of the differenced series (**y**); (d) correlogram of the residuals after fitting an AR(1) model (**z**).

8.4 Assessing evidence of long-term dependence

8.4.1 Nile minima

The data in the file `Nilemin.txt` are annual minimum water levels (mm) of the Nile River for the years 622 to 1284, measured at the Roda Island gauge near Cairo. It is likely that there may be a trend over a 600-year period due to changing climatic conditions or changes to the channels around Roda Island. We start the analysis by estimating and removing a linear trend fitted by regression. Having done this, a choice of `nar` is taken as a starting value for using `fracdiff` on the residuals from the regression. Given the iterative nature of the fitting process, the choice of initial values for `nar` and `nma` should not be critical. The estimate of d with `nar` set at 5 is 0.3457. The best-fitting model to the fractionally differenced series is AR(1) with parameter 0.14. We now re-estimate d using `fracdiff` with `nar` equal to 1, but in this case the estimate of d is unchanged. The residuals are a plausible realisation of DWN. The acf of the squared residuals indicates that a GARCH model would be appropriate. There is convincing evidence of long-term memory in the Nile River minima flows (Fig. 8.3).

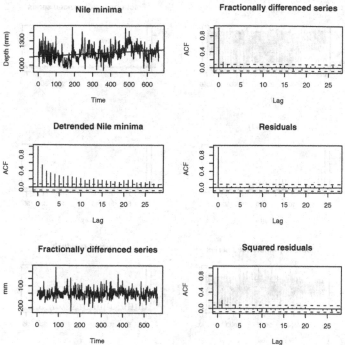

Fig. 8.3. Nile River minimum water levels: time series (top left); acf of detrended time series (middle left); fractionally differenced detrended series (lower left); acf of fractionally differenced series (top right); acf of residuals of AR(1) fitted to fractionally differenced series (middle right); acf of squared residuals of AR(1) (lower right).

8.4.2 Bellcore Ethernet data

The data in LAN.txt are the numbers of packet arrivals (bits) in 4000 consecutive 10-ms intervals seen on an Ethernet at the Bellcore Morristown Research and Engineering facility. A histogram of the numbers of bits is remarkably skewed, so we work with the logarithm of one plus the number of bits. The addition of 1 is needed because there are many intervals in which no packets arrive. The correlogram of this transformed time series suggests that a FARIMA model may be suitable.

The estimate of d, with nar set at 48, is 0.3405, and the fractionally differenced series has no substantial correlations. Nevertheless, the function ar fits an AR(26) model to this series, and the estimate of the standard deviation of the errors, 2.10, is slightly less than the standard deviation of the fractionally differenced series, 2.13. There is noticeable autocorrelation in the series of squared residuals from the AR(26) model, which is a feature of time series that have bursts of activity, and this can be modelled as a GARCH

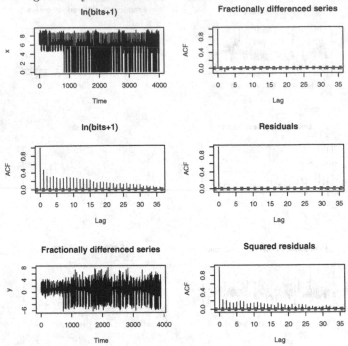

Fig. 8.4. Bellcore local area network (LAN) traffic, ln(1+number of bits): time series (top left); acf of time series (middle left); fractionally differenced series (lower left); acf of fractionally differenced series (top right); acf of residuals of AR(26) fitted to fractionally differenced series (middle right); acf of squared residuals of AR(26) (lower right).

process (Fig. 8.4). In Exercises 1 and 2, you are asked to look at this case in more detail and, in particular, investigate whether an ARMA model is more parsimonious.

8.4.3 Bank loan rate

The data in `mprime.txt` are of the monthly percentage US Federal Reserve Bank prime loan rate,[2] courtesy of the Board of Governors of the Federal Reserve System, from January 1949 until November 2007. The time series is plotted in the top left of Figure 8.5 and looks as though it could be a realisation of a random walk. It also has a period of high variability. The correlogram shows very high correlations at smaller lags and substantial correlation up to lag 28. Neither a random walk nor a trend is a suitable model for long-term

---

[2] Data downloaded from Federal Reserve Economic Data at the Federal Reserve Bank of St. Louis.

simulation of interest rates in a stable economy. Instead, we fit a FARIMA model, which has the advantage of being stationary.

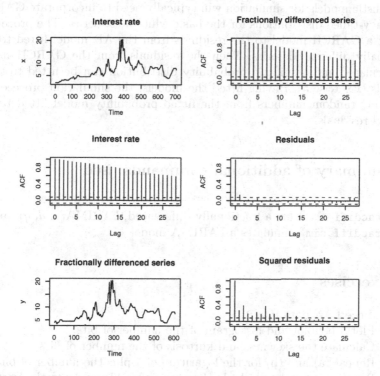

Fig. 8.5. Federal Reserve Bank interest rates: time series (top left); acf of time series (middle left); fractionally differenced series (lower left); acf of fractionally differenced series (upper right); acf of residuals of AR(17) fitted to fractionally differenced series (middle right); acf of squared residuals of AR(17) (lower right).

The estimate of d is almost 0, and this implies that the decay of the correlations from an initial high value is more rapid than it would be for a FARIMA model. The fitted AR model has an order of 17 and is not entirely satisfactory because of the statistically significant autocorrelation at lag 1 in the residual series. You are asked to do better in Exercise 3. The substantial autocorrelations of the squared residuals from the AR(17) model indicate that a GARCH model is needed. This has been a common feature of all three time series considered in this section.

8.5 Simulation

FARIMA models are important for simulation because short-memory models, which ignore evidence of long-memory, can lead to serious overestimation of system performance. This has been demonstrated convincingly at scales from reservoirs to routers in telecommunication networks.

Realistic models for simulation will typically need to incorporate GARCH and heavy-tailed distributions for the basic white noise series. The procedure is to fit a GARCH model to the residuals from the AR model fitted to the fractionally differenced series. Then the residuals from the GARCH model are calculated and a suitable probability distribution can be fitted to these residuals (Exercise 5). Having fitted the models, the simulation proceeds by generating random numbers from the fitted probability model fitted to the GARCH residuals.

8.6 Summary of additional commands used

`fracdiff`   fits a fractionally differenced, FARIMA(p, d, q), model
`fracdiff.sim` simulates a FARIMA model

8.7 Exercises

1. Read the LAN data into R.
   a) Plot a boxplot and histogram of the number of bits.
   b) Calculate the skewness and kurtosis of the number of bits.
   c) Repeat (a) and (b) for the logarithm of 1 plus the number of bits.
   d) Repeat (a) for the residuals after fitting an AR model to the fractionally differenced series.
   e) Fit an ARMA(p, q) model to the fractionally differenced series. Is this an improvement on the AR(p) model?
   f) In the text, we set `nar` in `fracdiff` at 48. Repeat the analysis with `nar` equal to 2.

2. Read the LAN data into R.
   a) Calculate the number of bits in 20-ms intervals, and repeat the analysis using this time series.
   b) Calculate the number of bits in 40-ms intervals, and repeat the analysis using this time series.
   c) Repeat (a) and (b) for realisations from FARIMA(0, d, 0).

3. Read the Federal Reserve Bank data into R.
   a) Fit a random walk model and comment.
   b) Fit an ARMA(p, q) model and comment.

4. The rescaled adjusted range is calculated for a time series $\{x_t\}$ of length m as follows. First compute the mean, $\bar{x}$, and standard deviation, s, of the series. Then calculate the adjusted partial sums

$$S_k = \sum_{t=1}^{k} x_t - k\bar{x}$$

for $k = 1, \ldots, m$. Notice that $S(m)$ must equal zero and that large deviations from 0 are indicative of persistence. The rescaled adjusted range

$$R_m = \{\texttt{max}(S_1, \ldots, S_m) - \texttt{min}(S_1, \ldots, S_m)\}/s$$

is the difference between the largest surplus and the greatest deficit. If we have a long time series of length n, we can calculate R_m for values of m from, for example, 20 upwards to n in steps of 10. When m is less than n, we can calculate $n - m$ values for R_m by starting at different points in the series. Hurst plotted $\ln(R_m)$ against $\ln(m)$ for many long time series. He noticed that lines fitted through the points were usually steeper for geophysical series, such as streamflow, than for realisations of independent Gaussian variables (Gaussian DWN). The average value of the slope (H) of these lines for the geophysical time series was 0.73, significantly higher than the average slope of 0.5 for the independent sequences. The linear logarithmic relationship is equivalent to

$$R_m \propto m^H$$

Plot $\ln(R_m)$ against $\ln(m)$ for the detrended Nile River minimum flows.

5. a) Refer to the data in LAN.txt and the time series of logarithms of the numbers of packet arrivals, with 1 added, in 10-ms intervals calculated from the numbers of packet arrivals. Fit a GARCH model to the residuals from the AR(26) model fitted to the fractionally differenced time series.

    b) Calculate the residuals from the GARCH model, and fit a suitable distribution to these residuals.

    c) Calculate the mean number of packets arriving in 10-ms intervals. Set up a simulation model for a router that has a realisation of the model in (a) as input and can send out packets at a constant rate equal to the product of the mean number of packets arriving in 10-ms intervals with a factor g, which is greater than 1.

    d) Code the model fitted in (a) so that it will provide simulations of time series of the number of packets that are the input to the router. Remember that you first obtain a realisation for ln(number of packets + 1) and then take the exponential of this quantity, subtract 1, and round the result to the nearest integer.

e) Compare the results of your simulation with a model that assumes Gaussian white noise for the residuals of the AR(26) model for $g = 1.05, 1.1, 1.5,$ and 2.

9

Spectral Analysis

9.1 Purpose

Although it follows from the definition of stationarity that a stationary time series model cannot have components at specific frequencies, it can nevertheless be described in terms of an average frequency composition. Spectral analysis distributes the variance of a time series over frequency, and there are many applications. It can be used to characterise wind and wave forces, which appear random but have a frequency range over which most of the power is concentrated. The British Standard BS6841, "Measurement and evaluation of human exposure to whole-body vibration", uses spectral analysis to quantify exposure of personnel to vibration and repeated shocks. Many of the early applications of spectral analysis were of economic time series, and there has been recent interest in using spectral methods for economic dynamics analysis (Iacobucci and Noullez, 2005).

More generally, spectral analysis can be used to detect periodic signals that are corrupted by noise. For example, spectral analysis of vibration signals from machinery such as turbines and gearboxes is used to expose faults before they cause catastrophic failure. The warning is given by the emergence of new peaks in the spectrum. Astronomers use spectral analysis to measure the red shift and hence deduce the speeds of galaxies relative to our own.

9.2 Periodic signals

9.2.1 Sine waves

Any signal that has a repeating pattern is periodic, with a period equal to the length of the pattern. However, the fundamental periodic signal in mathematics is the sine wave. Joseph Fourier (1768–1830) showed that sums of sine waves can provide good approximations to most periodic signals, and spectral analysis is based on sine waves.

P.S.P. Cowpertwait and A.V. Metcalfe, *Introductory Time Series with R*,
Use R, DOI 10.1007/978-0-387-88698-5_9,
© Springer Science+Business Media, LLC 2009

Spectral analysis can be confusing because different authors use different notation. For example, frequency can be given in radians or cycles per sampling interval, and frequency can be treated as positive or negative, or just positive. You need to be familiar with the sine wave defined with respect to a unit circle, and this relationship is so fundamental that the sine and cosine functions are called *circular functions*.

Imagine a circle with unit radius and centre at the origin, O, with the radius rotating at a rotational velocity of ω radians per unit of time. Let t be time. The angle, ωt, in radians is measured as the distance around the circumference from the positive real (horizontal) axis, with the anti-clockwise rotation defined as positive (Fig. 9.1). So, if the radius sweeps out a full circle, it has been rotated through an angle of 2π radians. The time taken for this one revolution, or *cycle*, is $2\pi/\omega$ and is known as the *period*.

The sine function, $\sin(\omega t)$, is the projection of the radius onto the vertical axis, and the cosine function, $\cos(\omega t)$, is the projection of the radius onto the horizontal axis. In general, a sine wave of *frequency* ω, *amplitude* A, and *phase* ψ is

$$A\sin(\omega t + \psi) \tag{9.1}$$

The positive phase shift represents an advance of $\psi/2\pi$ cycles. In spectral analysis, it is convenient to refer to specific sine waves as *harmonics*. We rely on the trigonometric identity that expresses a general sine wave as a weighted sum of sine and cosine functions:

$$A\sin(\omega t + \psi) = A\cos(\psi)\sin(\omega t) + A\sin(\psi)\cos(\omega t) \tag{9.2}$$

Equation (9.2) is fundamental for spectral analysis because a sampled sine wave of any given amplitude and phase can be fitted by a linear regression model with the sine and cosine functions as predictor variables.

9.2.2 Unit of measurement of frequency

The SI[1] unit of frequency is the hertz (Hz), which is 1 cycle per second and equivalent to 2π radians per second. The hertz is a derived SI unit, and in terms of fundamental SI units it has unit s^{-1}. A frequency of f cycles per second is equivalent to ω radians per second, where

$$\omega = 2\pi f \qquad \Leftrightarrow \qquad f = \frac{\omega}{2\pi} \tag{9.3}$$

The mathematics is naturally expressed in radians, but Hz is generally used in physical applications. By default, R plots have a frequency axis calibrated in cycles per sampling interval.

---

[1] SI is the International System of Units, abbreviated from the French *Le Systéme International d'Unités*.

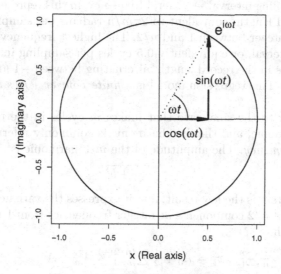

Fig. 9.1. Angle ωt is the length along the radius. The projection of the radius onto the x and y axes is $\cos(\omega t)$ and $\sin(\omega t)$, respectively.

9.3 Spectrum

9.3.1 Fitting sine waves

Suppose we have a time series of length n, $\{x_t : t = 1, \ldots, n\}$, where it is convenient to arrange that n is even, if necessary by dropping the first or last term. We can fit a time series regression with x_t as the response and $n - 1$ predictor variables:

$$\cos\left(\frac{2\pi t}{n}\right), \sin\left(\frac{2\pi t}{n}\right), \cos\left(\frac{4\pi t}{n}\right), \sin\left(\frac{4\pi t}{n}\right), \cos\left(\frac{6\pi t}{n}\right), \sin\left(\frac{6\pi t}{n}\right), \ldots,$$

$$\cos\left(\frac{2(n/2-1)\pi t}{n}\right), \sin\left(\frac{2(n/2-1)\pi t}{n}\right), \cos(\pi t).$$

We will denote the estimated coefficients by $a_1, b_1, a_2, b_2, a_3, b_3, \ldots, a_{n/2-1}$, $b_{n/2-1}, a_{n/2}$, respectively, so

$$x_t = a_0 + a_1\cos\left(\frac{2\pi t}{n}\right) + b_1\sin\left(\frac{2\pi t}{n}\right) + \cdots$$

$$+ a_{n/2-1}\cos\left(\frac{2(n/2 - 1)\pi t}{n}\right) + b_{n/2-1}\sin\left(\frac{2(n/2 - 1)\pi t}{n}\right) + a_{n/2}\cos(\pi t)$$

Since the number of coefficients equals the length of the time series, there are no degrees of freedom for error. The intercept term, a_0, is just the mean $\bar{x}$. The lowest frequency is one cycle, or 2π radians, per record length, which is $2\pi/n$

radians per sampling interval. A general frequency, in this representation, is m cycles per record length, equivalent to $2\pi m/n$ radians per sampling interval, where m is an integer between 1 and $n/2$. The highest frequency is π radians per sampling interval, or equivalently 0.5 cycles per sampling interval, and it makes $n/2$ cycles in the record length, alternating between -1 and $+1$ at the sampling points. This regression model is a *finite Fourier series* for a discrete time series.[2]

We will refer to the sine wave that makes m cycles in the record length as the mth harmonic, and the first harmonic is commonly referred to as the *fundamental frequency*. The amplitude of the mth harmonic is

$$A_m = \sqrt{a_m^2 + b_m^2}$$

Parseval's Theorem is the key result, and it expresses the variance of the time series as a sum of $n/2$ components at integer frequencies from 1 to $n/2$ cycles per record length:

$$\frac{1}{n}\sum_{t=1}^{n} x_t^2 = A_0^2 + \frac{1}{2}\sum_{m=1}^{(n/2)-1} A_m^2 + A_{n/2}^2$$

$$\mathrm{Var}(x) = \frac{1}{2}\sum_{m=1}^{(n/2)-1} A_m^2 + A_{n/2}^2 \tag{9.4}$$

Parseval's Theorem follows from the fact that the sine and cosine terms used as explanatory terms in the time series regression are uncorrelated, together with the result for the variance of a linear combination of variables (Exercise 1). A summary of the harmonics, and their corresponding frequencies and periods,[3] follows:

harmonic	period	frequency (cycle/samp. int.)	frequency (rad/samp. int.)	contribution to variance
1	n	$1/n$	$2\pi/n$	$\frac{1}{2}A_1^2$
2	$n/2$	$2/n$	$4\pi/n$	$\frac{1}{2}A_2^2$
3	$n/3$	$3/n$	$6\pi/n$	$\frac{1}{2}A_3^2$
$\vdots$	$\vdots$	$\vdots$	$\vdots$	$\vdots$
$n/2-1$	$n/(n/2-1)$	$(n/2-1)/n$	$(n-2)\pi/n$	$\frac{1}{2}A_{n/2-1}^2$
$n/2$	2	$1/n$	π	$A_{n/2}^2$

Although we have introduced the A_m in the context of a time series regression, the calculations are usually performed with the *fast fourier transform* algorithm (FFT). We say more about this in §9.7.

---

[2] A *Fourier series* is an approximation to a signal defined for continuous time over a finite period. The signal may have discontinuities. The Fourier series is the sum of an infinite number of sine and cosine terms.

[3] The period of a sine wave is the time taken for 1 cycle and is the reciprocal of the frequency measured in cycles per time unit.

9.3.2 Sample spectrum

A plot of A_m^2, as spikes, against m is a *Fourier line spectrum*. The *raw periodogram* in R is obtained by joining the tips of the spikes in the Fourier line spectrum to give a continuous plot and scaling it so that the area equals the variance. The periodogram distributes the variance over frequency, but it has two drawbacks. The first is that the precise set of frequencies is arbitrary inasmuch as it depends on the record length. The second is that the periodogram does not become smoother as the length of the time series increases but just includes more spikes packed closer together. The remedy is to smooth the periodogram by taking a moving average of spikes before joining the tips. The smoothed periodogram is also known as the *(sample) spectrum*. We denote the spectrum of $\{x_t\}$ by $C_{xx}()$, with an argument ω or f depending on whether it is expressed in radians or cycles per sampling interval. However, the smoothing will reduce the heights of peaks, and excessive smoothing will blur the features we are looking for. It is a good idea to consider spectra with different amounts of smoothing, and this is made easy for us with the R function spectrum. The argument span is the number of spikes in the moving average,[4] and is a useful guide for an initial value, for time series of lengths up to a thousand, is twice the record length.

The time series should either be mean adjusted (mean subtracted) before calculating the periodogram or the a_0 spike should be set to 0 before averaging spikes to avoid increasing the low-frequency contributions to the variance. In R, the spectrum function goes further than this and removes a linear trend from the series before calculating the periodogram. It seems appropriate to fit a trend and remove it if the existence of a trend in the underlying stochastic process is plausible. Although this will usually pertain, there may be cases in which you wish to attribute an apparent trend in a time series to a fractionally differenced process, and prefer not to remove a fitted trend. You could then use the fft function and average the spikes to obtain a spectrum of the unadjusted time series (§9.7).

The spectrum does not retain the phase information, though in the case of stationary time series all phases are equally likely and the sample phases have no theoretical interest.

9.4 Spectra of simulated series

9.4.1 White noise

We will start by generating an independent random sample from a normal distribution. This is a realisation of a Gaussian white noise process. If no span is specified in the spectrum function, R will use the heights of the Fourier line

---

[4] Weighted moving averages can be used, and the choice of weights determines the *spectral window*.

spectrum spikes to construct a spectrum with no smoothing.[5] We compare
this with a span of 65 in Figure 9.2.

```
> layout(1:2)
> set.seed(1)
> x <- rnorm(2048)
> spectrum(x, log = c("no"))
> spectrum(x, span = 65, log = c("no"))
```

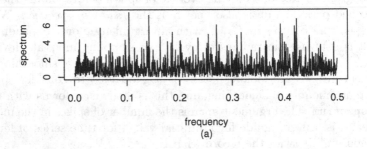

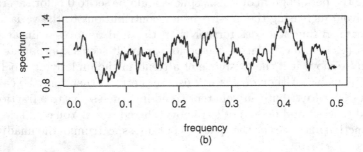

Fig. 9.2. Realisation of Gaussian white noise: (a) raw periodogram; (b) spectrum
with **span** = 65.

The default is a logarithmic scale for the spectrum, but we have changed
this by setting the **log** parameter to **"no"**. The frequency axis is cycles per
sampling interval.

The second spectrum is much smoother as a result of the moving average
of 65 adjacent spikes. Both spectra are scaled so that their area is one-half
the variance of the time series. The rationale for this is that the spectrum is

---

[5] By default, **spectrum** applies a taper to the first 10% and last 10% of the series and
pads the series to a highly composite length. However, 2048 is highly composite,
and the taper has little effect on a realisation of this length.

defined from -0.5 to 0.5, and is symmetric about 0. However, in the context of spectral analysis, there is no useful distinction between positive and negative frequencies, and it is usual to plot the spectrum over $[0, 0.5]$, scaled so that its area equals the variance of the signal. So, for a report it is better to multiply the R spectrum by a factor of 2 and to use hertz rather than cycles per sampling interval for frequency. You can easily do this with the following R commands, assuming the width of the sampling interval is `Del` (which would need to be assigned first):

```
> x.spec <- spectrum (x, span = 65, log = c("no"))
> spx <- x.spec$freq / Del
> spy <- 2 * x.spec$spec
> plot (spx, spy, xlab = "Hz", ylab = "variance/Hz", type = "l")
```

The theoretical spectrum for independent random variation with variance of unity is flat at 2 over the range $[0, 0.5]$. The name *white noise* is chosen to be reminiscent of white light made up from equal contributions of energy across the visible spectrum. An explanation for the flat spectrum arises from the regression model. If we have independent random errors, the $E[a_m]$ and $E[b_m]$ will all be 0 and the $E[A_m^2]$ are all equal. Notice that the vertical scale for the smoothed periodogram is from 0.8 to 1.4, so it is relatively flat (Fig. 9.2). If longer realisations are generated and the bandwidth is held constant, the default R spectra will tend towards a flat line at a height of 1.

The bandwidths shown in Figure 9.2 are calculated from the R definition of bandwidth as $\text{span} \times \{0.5/(n/2)\}/\sqrt{12}$. A more common definition of bandwidth in the context of spectral analysis is $\text{span}/(n/2)$ cycles per sampling interval. The latter definition is the spacing between statistically independent estimates of the spectrum height, and it is larger than the R bandwidth by a factor of 6.92.

The spectrum distributes variance over frequency, and the expected shape does not depend on the distribution that is being sampled. You are asked to investigate the effect, if any, of using random numbers from an exponential, rather than normal, distribution in Exercise 2.

9.4.2 AR(1): Positive coefficient

We generate a realisation of length 1024 from an AR(1) process with α equal to 0.9 and compare the time series plot, correlogram, and spectrum in Figure 9.3.

```
> set.seed(1)
> x <- w <- rnorm(1024)
> for (t in 2:1024) x[t]<- 0.9 * x[t-1] + w[t]
> layout(1:3)
> plot(as.ts(x))
> acf(x)
> spectrum(x, span = 51, log = c("no"))
```

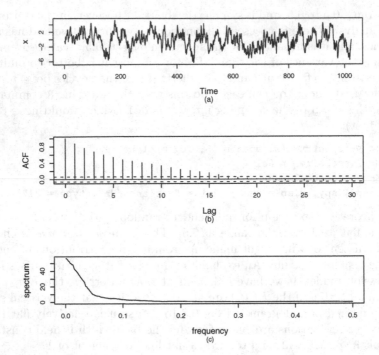

Fig. 9.3. Simulated AR(1) process with $\alpha = 0.9$: (a) time plot; (b) correlogram; (c) spectrum.

The plot of the time series shows the tendency for consecutive values to be relatively similar, and change is relatively slow, so we might expect the spectrum to pick up low-frequency variation. The acf quantifies the tendency for consecutive values to be relatively similar. The spectrum confirms that low-frequency variation dominates.

9.4.3 AR(1): Negative coefficient

We now change α from 0.9 to -0.9. The plot of the time series (Fig. 9.4) shows the tendency for consecutive values to oscillate, change is rapid, and we expect the spectrum to pick up high-frequency variation. The acf quantifies the tendency for consecutive values to oscillate, and the spectrum shows high frequency variation.

9.4.4 AR(2)

Consider an AR(2) process with parameters 1 and -0.6. This can be interpreted as a second-order difference equation describing the motion of a lightly damped single mode system (Exercise 3), such as a mass on a spring, subjected

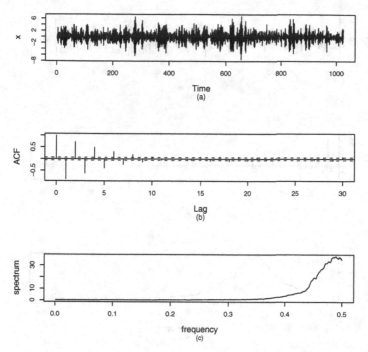

Fig. 9.4. Simulated AR(1) process with $\alpha = -0.9$: (a) time plot; (b) correlogram; (c) spectrum.

to a sequence of white noise impulses. The spectrum in Figure 9.5 shows a peak at the natural frequency of the system – the frequency at which the mass will oscillate if the spring is extended and then released.

```
> set.seed(1)
> x <- w <- rnorm(1024)
> for (t in 3:1024) x[t] <- x[t-1] - 0.6 * x[t-2] + w[t]
> layout (1:3)
> plot (as.ts(x))
> acf (x)
> spectrum (x, span = 51, log = c("no"))
```

9.5 Sampling interval and record length

Many time series are of an inherently continuous variable that is sampled to give a time series at discrete time steps. For example, the National Climatic Data Center (NCDC) provides 1-minute readings of temperature, wind speed, and pressure at meteorological stations throughout the United States. It is

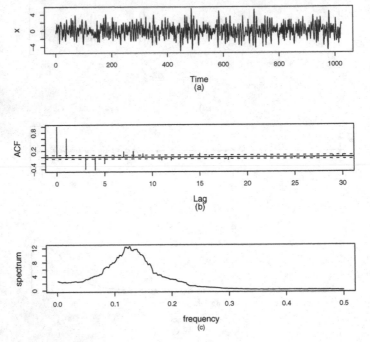

Fig. 9.5. Simulated AR(2) process with $\alpha_1 = 1$ and $\alpha_2 = -0.6$: (a) time plot; (b) correlogram; (c) spectrum.

crucial that the continuous signal be sampled at a sufficiently high rate to retain all its information. If the sampling rate is too low, we not only lose information but will mistake high-frequency variation for variation at a lower frequency. This latter phenomenon is known as *aliasing* and can have serious consequences.

In signal processing applications, the measurement device may return a voltage as a continuously varying electrical signal. However, analysis is usually performed on a digital computer, and the signal has to be sampled to give a time series at discrete time steps. The sampling is known as analog-to-digital conversion (A/D). Modern oscilloscopes sample at rates as high as Giga samples per second (GS/s) and have anti-alias filters, built from electronic components, that remove any higher-frequency components in the original continuous signal. Digital recordings of musical performances are typically sampled at rates of 1 Mega sample per second (MS/s) after any higher-frequencies have been removed with anti-alias filters. Since the frequency range of human hearing is from about 15 to 20,000 Hz, sampling rates of 1 MS/s are quite adequate for high-fidelity recordings.

9.5.1 Nyquist frequency

The Nyquist frequency is the cutoff frequency associated with a given sampling rate and is one-half the sampling frequency. Once a continuous signal is sampled, any frequency higher than the Nyquist frequency will be indistinguishable from its low-frequency alias.

To understand this phenomenon, suppose the sampling interval is Δ and the corresponding sampling frequency is $1/\Delta$ samples per second. A sine wave with a frequency of $1/\Delta$ cycles per second is generated by the radius in Figure 9.1 rotating anti-clockwise at a rate of 1 revolution per sampling interval Δ, and it follows that it cannot be detected when sampled at this rate. Similarly, a sine wave with a frequency of $-1/\Delta$ cycles per second, generated by the radius in Figure 9.1 rotating clockwise at a rate of 1 revolution per sampling interval Δ, is also undetectable. Now consider a sine wave with a frequency f that lies within the interval $[-1/(2\Delta), 1/(2\Delta)]$. This sine wave will be indistinguishable from any sine wave generated by a radius that completes an integer number of additional revolutions, anti-clockwise or clockwise, during the sampling interval. More formally, the frequency f will be indistinguishable from

$$f \pm k\Delta \qquad (9.5)$$

where k is an integer. Figure 9.6 shows a sine function with a frequency of 1 Hz, $\sin(2\pi t)$, sampled at 0.2 s, together with its alias when k in Equation (9.5) equals -1. This alias frequency is $1 - 1/0.2$, which equals -4 Hz. Physically, a frequency of -4 Hz is identical to a frequency of 4 Hz, except for a phase difference of half a cycle $(\sin(-\theta) = -\sin(\theta) = \sin(\theta - \pi))$.

```
> t  <- (0:10) / 5
> tc <- (0:2000) / 1000
> x  <- sin (2 * pi * t)
> xc <- sin (2 * pi * tc)
> xa <- sin (-4 * 2 * pi * tc)
> plot (t, x)
> lines (tc, xc)
> lines (tc, xa, lty = "dashed")
```

To summarise, the Nyquist frequency Q is related to the sampling interval Δ by

$$Q = \frac{1}{2\Delta} \qquad (9.6)$$

and Q should be higher than any frequency components in the continuous signal.

9.5.2 Record length

To begin with, we need to establish the highest frequency we can expect to encounter and set the Nyquist frequency Q well above this. The Nyquist frequency determines the sampling interval, Δ, from Equation (9.6). If the time

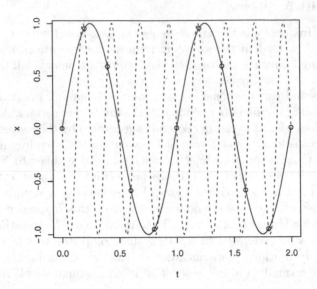

Fig. 9.6. Aliased frequencies: 1 Hz and 4 Hz with $\Delta = 0.2$ second.

series has length n, the record length, T, is $n\Delta$. The fundamental frequency is $1/T$ Hz, and this is the spacing between spikes in the Fourier line spectrum. If we wish to distinguish frequencies separated by ϵ Hz, we should aim for independent estimates of the spectrum centred on these frequencies. This implies that the bandwidth must be at most ϵ. If we take a moving average of L spikes in the Fourier line spectrum, we have the following relationship:

$$\frac{2L}{n\Delta} = \frac{2L}{T} \leq \epsilon \tag{9.7}$$

For example, suppose we wish to distinguish frequencies separated by 1 Hz in an audio recording. A typical sampling rate for audio recording is 1 MS/s, corresponding to $\Delta = 0.000001$. If we take L equal to 100, it follows from Equation (9.7) that n must exceed 200×10^6. This is a long time series but the record length is less than four minutes. If a time series of this length presents computational problems, an alternative method for computing a smoothed spectrum is to calculate the Fourier line spectrum for the 100 subseries of two million observations and average these 100 Fourier line spectra.

9.6 Applications

9.6.1 Wave tank data

The data in the file wave.dat are the surface height, relative to still water level, of water at the centre of a wave tank sampled over 39.6 seconds at a rate of 10 samples per second. The aim of the analysis is to check whether the spectrum is a realistic emulation of typical sea spectra. Referring to Figure 9.7, the time series plot gives a general impression of the wave profile over time and we can see that there are no obvious erroneous values. The correlogram is qualitatively similar to that for a realisation of an AR(2) process,[6] but an AR(2) model would not account for a second peak in the spectrum at a frequency near 0.09.

```
> www <- "http://www.massey.ac.nz/~pscowper/ts/wave.dat"
> wavetank.dat <- read.table(www, header=T)
> attach (wavetank.dat)
> layout (1:3)
> plot (as.ts(waveht))
> acf (waveht)
> spectrum (waveht)
```

The default method of fitting the spectrum used above does not require the ar function. However, the ar function is used in §9.9 and selects an AR(13) model. The shape of the estimated spectrum in Figure 9.7 is similar to that of typical sea spectra.

9.6.2 Fault detection on electric motors

Induction motors are widely used in industry, and although they are generally reliable, they do require maintenance. A common fault is broken rotor bars, which reduce the output torque capability and increase vibration, and if left undetected can lead to catastrophic failure of the electric motor. The measured current spectrum of a typical motor in good condition will have a spike at mains frequency, commonly 50 Hz, with side band peaks at 46 Hz and 54 Hz. If a rotor bar breaks, the magnitude of the side band peaks will increase by a factor of around 10. This increase can easily be detected in the spectrum.

Siau et al. (2004) compare current spectra for an induction motor in good condition and with one broken bar. They sample the current at 0.0025-second intervals, corresponding to a Nyquist frequency of 200 Hz, and calculate spectra from records of 100 seconds length. The time series have length 40,000, and the bandwidth with a span of 60 is 1.2 Hz (Equation (9.7)).

The data are in the file imotor.txt. R code for drawing the spectra (Fig. 9.8) follows. The broken bar condition is indicated clearly by the higher side band peaks in the spectrum. In contrast, the standard deviations of the good condition and broken condition time series are very close.

---

[6] The pacf, not shown here, also suggests that an AR(2) model would be plausible.

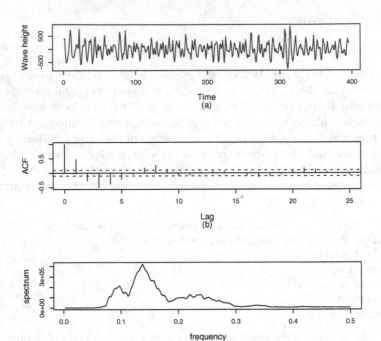

Fig. 9.7. Wave elevation series: (a) time plot; (b) correlogram; (c) spectrum.

```
> www <- "http://www.massey.ac.nz/~pscowper/ts/imotor.txt"
> imotor.dat <- read.table(www, header = T)
> attach (imotor.dat)
> xg.spec <- spectrum(good,    span = 9)
> xb.spec <- spectrum(broken, span = 9)
> freqg <- 400 * xg.spec$freq [4400:5600]
> freqb <- 400 * xb.spec$freq [4400:5600]
> plot(freqg, 10*log10(xg.spec$spec[4400:5600]), main = "",
    xlab = "Frequency (Hz)", ylab = "Current spectrum (dB)", type="l")
> lines(freqb, 10 * log10(xb.spec$spec[4400:5600]), lty = "dashed")
> sd(good)
[1] 7071.166
> sd(broken)
[1] 7071.191
```

9.6.3 Measurement of vibration dose

The drivers of excavators in open cast mines are exposed to considerable mechanical vibration. The British Standard Guide BS6841:1987 is routinely used to quantify the effects. A small engineering company has developed an active

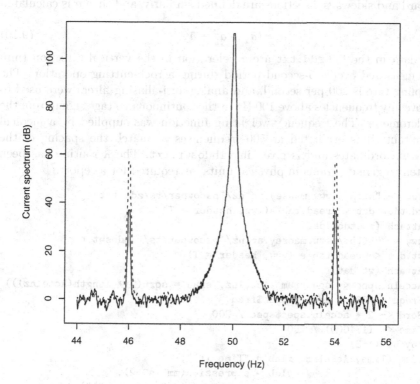

Fig. 9.8. Spectrum of current signal from induction motor in good condition (solid) and with broken rotor bar (dotted). Frequency is in cycles per 0.0025 second sampling interval.

vibration absorber for excavators and has carried out tests. The company has accelerometer measurements of the acceleration in the forward (x), sideways (y), and vertical (z) directions during a rock-cutting operation. The estimated vibration dose value is defined as

$$eVDV = \left[(1.4 \times \bar{a})^4 \times T\right]^{1/4} \qquad (9.8)$$

where $\bar{a}$ is the root mean square value of frequency-weighted acceleration (ms^{-2}) and T is the duration (s). The mean square frequency-weighted acceleration in the vertical direction is estimated by

$$\bar{a}_z^2 = \int C_{\ddot{z}\ddot{z}}(f)W(f)\,df \qquad (9.9)$$

where the weighting function, $W(f)$, represents the relative severity of vibration at different frequencies for a driver, and the acceleration time series is the second derivative of the displacement signal, denoted $\ddot{z}$. Components in the

forward and sideways directions are defined similarly, and then $\bar{a}$ is calculated
as

$$\bar{a} = (\bar{a}_x^2 + \bar{a}_y^2 + \bar{a}_z^2)^{1/2} \qquad (9.10)$$

The data in the file zdd.txt are acceleration in the vertical direction (mm s^{-2}) measured over a 5-second period during a rock-cutting operation. The sampling rate is 200 per second, and analog anti-aliasing filters were used to remove any frequencies above 100 Hz in the continuous voltage signal from the accelerometer. The frequency-weighting function was supplied by a medical consultant. It is evaluated at 500 frequencies to match the spacing of the spectrum ordinates and is given in vibdoswt.txt. The R routine has been written to give diagrams in physical units, as required for a report.[7]

```
> www <- "http://www.massey.ac.nz/~pscowper/ts/zdd.txt"
> zdotdot.dat <- read.table(www, header = T)
> attach (zdotdot.dat)
> www <- "http://www.massey.ac.nz/~pscowper/ts/vibdoswt.txt"
> wt.dat <- read.table (www, header = T)
> attach (wt.dat)
> acceln.spec <- spectrum (Accelnz, span = sqrt(2 * length(Accelnz)))
> Frequ <- 200 * acceln.spec$freq
> Sord <- 2 * acceln.spec$spec / 200
> Time <- (1:1000) / 200
> layout (1:3)
> plot (Time, Accelnz, xlab = "Time (s)",
                  ylab = expression(mm~ s^-2),
                  main = "Acceleration", type = "l")
> plot (Frequ, Sord, main = "Spectrum", xlab = "Frequency (Hz)",
                  ylab = expression(mm^2~s^-4~Hz^-1), type = "l")
> plot (Frequ, Weight, xlab = "Frequency (Hz)",
                  main = "Weighting function", type = "l")
> sd (Accelnz)
  [1] 234.487
> sqrt( sum(Sord * Weight) * 0.2 )
  [1] 179.9286
```

Suppose a driver is cutting rock for a 7-hour shift. The estimated root mean square value of frequency weighted acceleration is 179.9 (mm s^{-2}). If we assume continuous exposure throughout the 7-hour period, the eVDV calculated using Equation (9.8) is 3.17 (m s$^{-1.75}$). The British Standard states that doses as high as 15 will cause severe discomfort but is non-committal about safe doses arising from daily exposure. The company needs to record acceleration measurements during rock-cutting operations on different occasions, with and without the vibration absorber activated. It can then estimate the decrease in vibration dose that can be achieved by fitting the vibration absorber to an excavator (Fig. 9.9).

---

[7] Within R, type demo(plotmath) to see a list of mathematical operators that can be used by the function expression for plots.

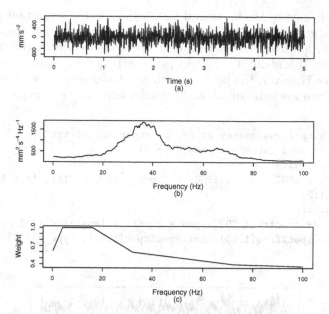

Fig. 9.9. Excavator series: (a) acceleration in vertical direction; (b) spectrum; (c) frequency weighting function.

9.6.4 Climatic indices

Climatic indices are strongly related to ocean currents, which have a major influence on weather patterns throughout the world. For example, El Niño is associated with droughts throughout much of eastern Australia. A statistical analysis of these indices is essential for two reasons. Firstly, it helps us assess evidence of climate change. Secondly, it allows us to forecast, albeit with limited confidence, potential natural disasters such as droughts and to take action to mitigate the effects. Farmers, in particular, will modify their plans for crop planting if drought is more likely than usual. Spectral analysis enables us to identify any tendencies towards periodicities or towards persistence in these indices.

The *Southern Oscillation Index* (SOI) is defined as the normalised pressure difference between Tahiti and Darwin. El Niño events occur when the SOI is strongly negative, and are associated with droughts in eastern Australia. The monthly time series[8] from January 1866 until December 2006 are in soi.txt. The time series plot in Figure 9.10 is a useful check that the data have been read correctly and gives a general impression of the range and variability of the SOI. But, it is hard to discern any frequency information. The spectrum is plotted with a logarithmic vertical scale and includes a 95% confidence interval for the population spectrum in the upper right. The confidence interval can be represented as a vertical line relative to the position of the sample

---

[8] More details and the data are at http://www.cru.uea.ac.uk/cru/data/soi.htm.

spectrum indicated by the horizontal line, because it has a constant width on a logarithmic scale (§9.10.2). The spectrum has a peak at a low-frequency, so we enlarge the low frequency section of the spectrum to identify this frequency more precisely. It is about 0.022 cycles per month and corresponds to a period of 45 months. However, the peak is small and lower frequency contributions to the spectrum are substantial, so we cannot expect a regular pattern of El Niño events.

```
> www <- "http://www.massey.ac.nz/~pscowper/ts/soi.txt"
> soi.dat <- read.table(www, header = T)
> attach (soi.dat)
> soi.ts <- ts(SOI, st = c(1866, 1), end = c(2006, 11), fr = 12)
> layout (1:3)
> plot (soi.ts)
> soi.spec <- spectrum( SOI, span = sqrt(2 * length(SOI)) )
> plot (soi.spec$freq[1:60], soi.spec$spec[1:60], type = "l")
```

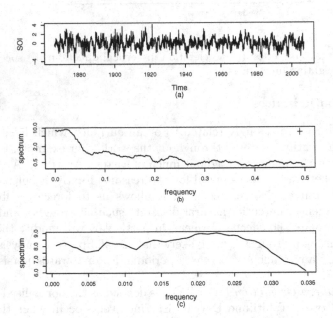

Fig. 9.10. Southern Oscillation Index: (a) time plot; (b) spectrum; (c) spectrum for the low-frequencies.

The *Pacific Decadal Oscillation* (PDO) index is the difference between an average of sea surface temperature anomalies in the North Pacific Ocean poleward of 20 °N and the monthly mean global average anomaly.[9] The monthly time series from January 1900 until November 2007 is in pdo.txt. The spectrum in Figure 9.11 has no noteworthy peak and increases as the frequency

---

[9] The time series data are available from http://jisao.washington.edu/pdo/.

becomes lower. The function `spectrum` removes a fitted linear trend before calculating the spectrum, so the increase as the frequency tends to zero is evidence of long-term memory in the PDO.

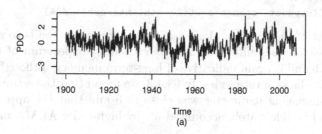

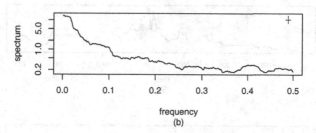

Fig. 9.11. Pacific Decadal Oscillation: (a) time plot; (b) spectrum.

```
> www <- "http://www.massey.ac.nz/~pscowper/ts/pdo.txt"
> pdo.dat <- read.table(www, header = T)
> attach (pdo.dat)
> pdo.ts <- ts( PDO, st = c(1900, 1), end = c(2007, 11), fr = 12 )
> layout (1:2)
> plot (pdo.ts)
> spectrum( PDO, span = sqrt( 2 * length(PDO) ) )
```

This analysis suggests that a FARIMA model might be suitable for modelling the PDO and for generating future climate scenarios.

9.6.5 Bank loan rate

The data in `mprime.txt` are the monthly percentage US Federal Reserve Bank prime loan rate,[10] courtesy of the Board of Governors of the Federal Reserve System, from January 1949 until November 2007. We will plot the time series, the correlogram, and a spectrum on a logarithmic scale (Fig. 9.12).

---

[10] Data downloaded from Federal Reserve Economic Data at the Federal Reserve Bank of St. Louis.

```
> www <- "http://www.massey.ac.nz/~pscowper/ts/mprime.txt"
> intr.dat <- read.table(www, header = T)
> attach (intr.dat)
> layout (1:3)
> plot (as.ts(Interest), ylab = 'Interest rate')
> acf (Interest)
> spectrum(Interest, span = sqrt(length(Interest)) / 4)
```

The height of the spectrum increases as the frequency tends to zero (Fig. 9.12). This feature is similar to that observed in the spectrum of the PDO series in §9.6.5 and is again indicative of long-term memory, although it is less pronounced in the loan rate series. In §8.4.3, we found that the estimate of the fractional differencing parameter was close to 0 and that the apparent long memory could be adequately accounted for by high-order ARMA models.

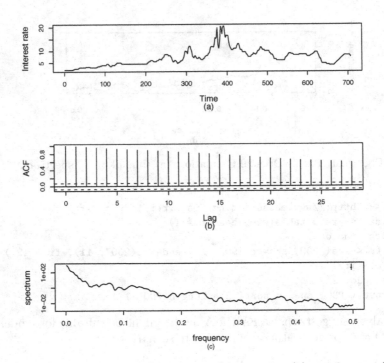

Fig. 9.12. Federal Reserve Bank loan rates: (a) time plot; (b) correlogram; (c) spectrum.

9.7 Discrete Fourier transform (DFT)*

The theoretical basis for spectral analysis can be described succinctly in terms of the discrete Fourier transform (DFT). The DFT requires the concept of complex numbers and Euler's formula for a complex sinusoid, but the theory then follows nicely. In R, complex numbers are handled by typing i following, without a space, a numerical value; for example,

```
> z1 <- 2 + (0+3i)
> z2 <- -1 - (0+1i)
> z1 - z2

[1] 3+4i

> z1 * z2

[1] 1-5i

> abs(z1)

[1] 3.61
```

Euler's formula for a complex sinusoid is

$$e^{i\theta} = \cos(\theta) + i\sin(\theta) \tag{9.11}$$

If the circle in Figure 9.1 is at the centre of the complex plane, $e^{i\theta}$ is the point along the circumference. This remarkable formula can be verified using Taylor expansions of $e^{i\theta}$, $\sin(\theta)$, and $\cos(\theta)$.

The DFT is usually calculated using the fast fourier transform algorithm (FFT), which is very efficient for long time series. The DFT of a time series of length n, $\{x_t : t = 0, \ldots, n-1\}$, and its inverse transform (IDFT) are defined by Equation (9.12) and Equation (9.13), respectively.

$$X_m = \sum_{t=0}^{n-1} x_t e^{-2\pi i m t/n} \qquad m = 0, \ldots, n-1 \tag{9.12}$$

$$x_t = \frac{1}{n} \sum_{m=0}^{n-1} X_m e^{2\pi i t m/n} \qquad t = 0, \ldots, n-1 \tag{9.13}$$

It is convenient to start the time series at $t = 0$ for these definitions because m then corresponds to frequency $2\pi m/n$ radians per sampling interval. The steps in the derivation of the DFT-IDFT transform pair are set out in Exercise 5. The DFT is obtained in R with the function fft(), where x[t+1] corresponds to x_t and X[m+1] corresponds to X_m.

```
> set.seed(1)
> n <- 8
> x <- rnorm(n)
> x

[1] -0.626  0.184 -0.836  1.595  0.330 -0.820  0.487  0.738

> X <- fft(x)
> X
```

```
[1]   1.052+0.000i -0.852+0.007i  0.051+2.970i -1.060-2.639i
[5]  -2.342+0.000i -1.060+2.639i  0.051-2.970i -0.852-0.007i
```

```
> fft(X, inverse = TRUE)/n
```

```
[1] -0.626-0i  0.184+0i -0.836-0i  1.595-0i  0.330+0i -0.820-0i
[7]  0.487+0i  0.738+0i
```

The complex form of Parseval's Theorem, first given in Equation (9.4), is

$$\sum_{t=0}^{n-1} x_t^2 = \sum_{m=0}^{n-1} |X_m|^2/n \tag{9.14}$$

If n is even, the $|X_m|^2$ contribution to the variance corresponds to a frequency of $2\pi m/n$ for $m = 1, \ldots, n/2$. For $m = n/2, \ldots, (n-1)$, the frequencies are greater than the Nyquist frequency, π, and are aliased to the frequencies $2\pi(m-n)/n$, which lie in the range $[-\pi, -2\pi/n]$. All but two of the X_m occur as complex conjugate pairs; that is, $X_{n-j} = X_j^*$ for $j = 1, \ldots, n/2 - 1$. The following lines of R code give the spikes of the Fourier line spectrum FL at frequencies in frq scaled so that FL[1] is mean(x)^2 and the sum of FL[2], ..., FL[n/2+1] is(n-1)*var(x)/n.

```
> fq <- 2 * pi / n
> frq <- 0
> FL  <- 0
> FL [1] <- X[1]^2 / n^2
> frq[1] <- 0
> for ( j in 2:(n/2) ) {
    FL [j] <- 2 * (X[j] * X[n+2-j]) / n^2
    frq[j] <- fq * (j-1)
  }
> FL [n/2 + 1] <- X[n/2 + 1]^2 / n^2
> frq[n/2 + 1] <- pi
```

If a plot is required, plot(frq,FL) can be used. You can now average spikes as you wish to obtain a spectrum (Exercise 5).

9.8 The spectrum of a random process*

Although we can now calculate the spectrum of a time series of finite length, we have no algebraic formula for defining the spectrum of the underlying random process. The definition of the spectrum of a random process follows from considering the expected value of a smoothed periodogram and is

$$\Gamma(\omega) = \frac{1}{2\pi} \sum_{k=-\infty}^{\infty} \gamma_k e^{-i\omega k} \qquad -\pi < \omega < \pi \tag{9.15}$$

The derivation of Equation (9.15) is given in §9.8.3.

9.8.1 Discrete white noise

The spectrum of discrete white noise with variance σ^2 is easily obtained from the definition since the only non-zero value of γ_k is σ^2 when $k = 0$.

$$\Gamma(\omega) = \frac{\sigma^2}{2\pi} \qquad -\pi < \omega < \pi \tag{9.16}$$

The area under the spectrum is the variance σ^2.

9.8.2 AR

The spectrum of an ARMA(p, q) process is

$$\Gamma(\omega) = \frac{\sigma^2}{2\pi} \left| \frac{1 + \sum_{l=1}^{q} \beta_l e^{-il\omega}}{1 + \sum_{j=1}^{p} \alpha_j e^{-ij\omega}} \right|^2 \qquad -\pi < \omega < \pi \tag{9.17}$$

It is far easier to derive Equation (9.17) from results we develop in Chapter 10, but we state the result here because it suggests another method of estimating the spectrum of a random process.

9.8.3 Derivation of spectrum

Assume that $\{x_t : t = 0, \ldots, n-1\}$ is a time series with mean 0. The contribution to the Fourier line spectrum at frequency m is

$$|X(m)|^2 = X_m^* X(m) = \sum_{t=0}^{n-1} x_t e^{2\pi i m t/n} \sum_{s=0}^{n-1} x_t e^{2\pi i m s/n} \tag{9.18}$$

which can be rewritten as

$$\sum_{t=0}^{n-1} \sum_{s=0}^{n-1} x_t x_s e^{2\pi i m(s-t)/n} \tag{9.19}$$

We now substitute $k = s - t$ and change the variables to t using k instead of s. Then the double sum becomes

$$\sum_{t=0}^{n-1} \sum_{k=-t}^{n-1-t} x_t x_{t+k} e^{-2\pi i m k/n} \tag{9.20}$$

Since the mean of the time series is 0, the sum of $x_t x_{t+k}$ is proportional to the sample autocovariance at lag k, so Equation (9.20) can be written as

$$\sum_{k=-(n-1)}^{n-1} c_k e^{-2\pi i m k/n} \tag{9.21}$$

This will not converge as n tends to infinity because as n increases we introduce more spikes into the spectrum. However, if we take the expected value of Equation (9.21), and let $n \to \infty$ so that

$$E\left[c_k\right] \to \gamma_k \tag{9.22}$$

and define

$$\lim_{n \to \infty} \frac{2\pi m}{n} = \omega \tag{9.23}$$

Equation (9.15) follows. The factor $1/2\pi$ in Equation (9.15) is a normalising factor that ensures that the area under the spectrum equals the variance of the stochastic process.

9.9 Autoregressive spectrum estimation

Another method for estimating the spectrum from a time series is to fit a suitable ARMA(p, q) model and then use Equation (9.17) to calculate the corresponding spectrum. It is usual to use a high order AR(p) model rather than the more general ARMA model, and this is an option with `spectrum` that is invoked by including `method=c("ar")` as an argument in the function. It gives a rather smooth estimate of the spectrum, increasingly so as p becomes smaller; it is used below on the wave tank data. The function determines a suitable order for the AR(p) model using the AIC; the `span` parameter is not needed.

```
> spectrum( waveht, log = c("no"), method = c("ar") )
```

The smooth shape is useful for qualitative comparisons with the sea spectra (Fig. 9.13). The analysis also indicates that we could use an AR(13) model to obtain realisations of time series with this same spectrum in computer simulations. A well-chosen probability distribution for the errors could be used to give a realistic simulation of extreme values in the series.

9.10 Finer details

9.10.1 Leakage

Suppose a time series is a sampled sine function at a specific frequency. If this frequency corresponds to one of the frequencies in the finite Fourier series, then there will be a spike in the Fourier line spectrum at this frequency. This coincidence is unlikely to arise by chance, so now suppose that the specific frequency lies between two of the frequencies in the finite Fourier series. There will not only be spikes at these two frequencies but also smaller spikes at neighbouring frequencies (Exercise 6). This phenomenon is known as *leakage*.

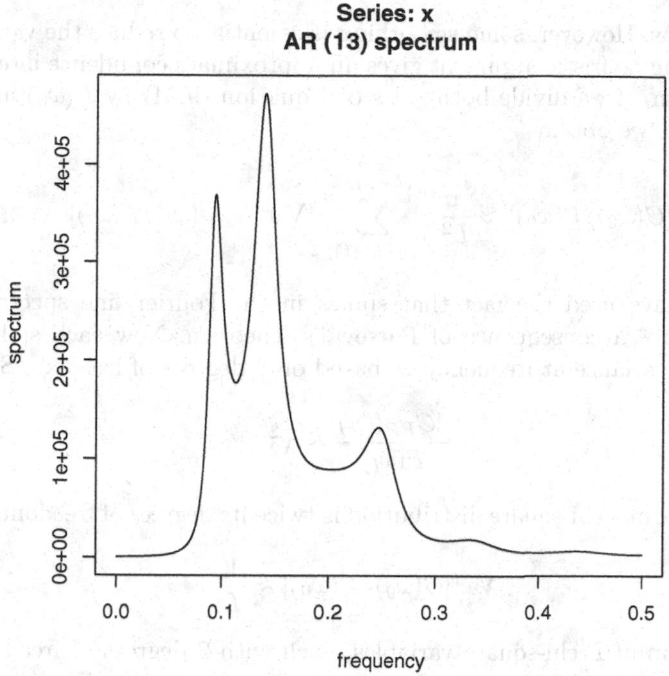

Series: x
AR (13) spectrum

Fig. 9.13. Wave elevation series: spectrum calculated from fitting an AR model.

9.10.2 Confidence intervals

Consider a frequency ω_0 corresponding to a spike of the Fourier line spectrum. If we average an odd number, L, of scaled spikes to obtain a smoothed spectrum, then

$$C(\omega_0) = \frac{1}{L} \sum_{l=-(L-1)/2}^{(L-1)/2} C_{RP}(\omega_l) \qquad (9.24)$$

where C_{RP} are the raw periodogram, scaled spike estimates. Now taking the expectation of both sides of Equation (9.24), and assuming the raw periodogram is unbiased for the population spectrum, we obtain

$$E\left[C(\omega_0)\right] = \frac{1}{L} \sum_{l=-(L-1)/2}^{(L-1)/2} \Gamma(\omega_l) \qquad (9.25)$$

Provided the population spectrum does not vary much over the interval $\left[-\omega_{-(L-1)/2}, \omega_{(L-1)/2}\right]$,

$$E\left[C(\omega_0)\right] \approx \Gamma(\omega_0) \qquad (9.26)$$

But, notice that if ω_0 corresponds to a peak or trough of the spectrum, the smoothed spectrum will be biased low or high. The more the smoothing, the more the bias. However, some smoothing is essential to reduce the variability. The following heuristic argument gives an approximate confidence interval for the spectrum. If we divide both sides of Equation (9.24) by $\Gamma(\omega_0)$ and take the variance, we obtain

$$\text{Var}\left[C(\omega_0)/\Gamma(\omega_0)\right] \approx \frac{1}{L^2} \sum_{l=-(L-1)/2}^{(L-1)/2} \text{Var}\left[C_{RP}(\omega_l)/\Gamma(\omega_l)\right] \qquad (9.27)$$

where we have used the fact that spikes in the Fourier line spectrum are independent – a consequence of Parseval's Theorem. Now each spike is an estimate of variance at frequency ω_l based on 2 degrees of freedom. So,

$$\frac{2C_{RP}(\omega_l)}{\Gamma(\omega_l)} \sim \chi_2^2 \qquad (9.28)$$

The variance of a chi-square distribution is twice its degrees of freedom. Hence,

$$\text{Var}\left[C(\omega_0)/\Gamma(\omega_0)\right] \approx \frac{1}{L} \qquad (9.29)$$

A scaled sum of L chi-square variables, each with 2 degrees of freedom, is a scaled chi-square variable with $2L$ degrees of freedom and well approximated by a normal distribution. Thus an approximate 95% confidence interval for $\Gamma(\omega)$ is

$$\left[\left(1 - \frac{2}{\sqrt{L}}\right) C(\omega), \left(1 + \frac{2}{\sqrt{L}}\right) C(\omega)\right] \qquad (9.30)$$

We have dropped the subscript on ω because the result remains a good approximation for estimates of the spectrum interpolated between $C(\omega_l)$.

9.10.3 Daniell windows

The function `spectrum` uses a modified Daniell window, or smoother, that gives half weight to the end values. If more than one number is specified for the parameter `span`, it will use a series of Daniell smoothers, and the net result will be a centred moving average with weights decreasing from the centre. The rationale for using a series of smoothers is that it will decrease the bias.

9.10.4 Padding

The simplest FFT algorithm assumes that the time series has a length that is some power of 2. A positive integer is highly composite if it has more divisors than any smaller positive integer. The FFT algorithm is most efficient when the length n is *highly composite*, and by default `spec.pgram` pads the mean adjusted time series with zeros to reach the smallest highly composite number that is greater than or equal to the length of the time series. Padding can be avoided by setting the parameter `fast=FALSE`. A justification for padding is that the length of the time series is arbitrary and that adding zeros has no effect on the frequency composition. Adding zeros does reduce the variance, and this must be remembered when scaling the spectrum, so that its area equals the variance of the original time series.

9.10.5 Tapering

The length of a time series is not usually related to any underlying frequency composition. However, the discrete Fourier series keeps replicating the original time series as $-\infty < t < \infty$, known as *periodic extension* of the original time series, and there will usually be a jump between the end of one replicate time series and the start of the next. These jumps can be avoided by reducing the magnitude of the values of the time series, relative to its mean, at the beginning and towards the end. The default with `spectrum` is a taper applied to 10% of the data at the beginning and towards the end of the time series. Tapering increases the variance of Fourier line spectrum spikes but reduces the bias (Exercise 7). It will also reduce the variance of the time series. The default proportion of data to which the taper is applied can be changed with the parameter `taper`. The `fft` function does not remove the mean, remove a linear trend, or apply a taper, operations that are generally classed as *pre-processing*.

9.10.6 Spectral analysis compared with wavelets

Spectral analysis is appropriate for the analysis of stationary time series and for identifying periodic signals that are corrupted by noise. Spectral analysis can be used for spatial series such as surface roughness transects, and two-dimensional spectral analysis can be used for measurements of surface roughness made over a plane. However, spectral analysis is not suitable for non-stationary applications.

In contrast, wavelets have been developed to summarise the variation in frequency composition through time or over space. There are many applications, including compression of digital files of images and in speech recognition software. Nason (2008) provides an introduction to wavelets using the R package `WaveThresh4`.

9.11 Summary of additional commands used

`spectrum`	returns the spectrum
`spec.pgam`	returns the spectrum with more control of parameters
`fft`	returns the DFT

9.12 Exercises

1. Refer to §9.3.1 and take $n = 128$.
   a) Use R to calculate $\cos(2\pi t/n)$, $\sin(2\pi t/n)$, and $\cos(4\pi t/n)$ for $t = 1, \ldots, n$. Calculate the three variances and the three correlations.
   b) Assuming the results above generalise, provide an explanation for Parseval's Theorem.
   c) Explain why the $A_{n/2}^2$ term in Equation (9.4) is not divided by 2.

2. Repeat the investigation of realisations from AR processes in §9.4 using random deviates from an exponential distribution with parameter 1 and with its mean subtracted, rather than the standard normal distribution.

3. The differential equation for the oscillatory response x of a lightly damped single mode of vibration system, such as a mass on a spring, with a forcing term w is
$$\ddot{x} + 2\zeta\Omega\dot{x} + \Omega^2 x = w$$
   where ζ is the damping coefficient, which must be less than 1 for an oscillatory response, and Ω is the natural frequency. Approximate the derivatives by backward differences:
$$\ddot{x} = x_t - 2x_{t-1} + x_{t-2} \qquad \dot{x} = x_t - x_{t-1}$$
   and set $w = w_t$ and rearrange to obtain the form of the AR(2) process in §8.4.4. Consider an approximation using central differences.

4. Suppose that
$$x_t = \sum_{m=0}^{n-1} a_m e^{2\pi imt/n} \qquad m = 0, \ldots, n-1 \tag{9.31}$$
   for some coefficients a_m that we wish to determine. Now multiply both sides of this equation by $e^{-2\pi ijt/n}$ and sum over t from 0 to $n-1$ to obtain
$$\sum_{t=0}^{n-1} x_t e^{-2\pi ijt/n} = \sum_{t=0}^{n-1}\sum_{m=0}^{n-1} a_m e^{2\pi i(m-j)t/n} \tag{9.32}$$
   Consider a fixed value of j. Notice that the sum to the right of a_m is a geometric series with sum 0 unless $m = j$. This is Equation (9.12) expressed it terms of na_j in place of X_m with a factor of n.

5. Write R code to average an odd number of spike heights obtained from fft and hence plot a spectrum.

6. Sample the three signals
   a) $\sin(\pi t/2)$
   b) $\sin(3\pi t/4)$
   c) $\sin(5\pi t/8)$

   at times $t = 0, \ldots, 7$, using `fft` to compare their line spectra.

7. Sample the signal $\sin(11\pi t/32)$ for $t = 0, \ldots, 31$. Use `fft` to calculate the Fourier line spectrum. The cosine bell taper applied to the beginning α and ending α of a series is defined by

$$\left[1 - \cos\left(\pi\{t + 0.5\}/\{\alpha n\}\right)\right] x_t \qquad (t+1) \le \alpha n$$

$$\left[1 - \cos\left(\pi\{n - t - 0.5\}/\{\alpha n\}\right)\right] x_t \qquad (t+1) \ge (1 - \alpha)n$$

   Investigate the effect of this taper, with $\alpha = 0.1$, on the Fourier line spectrum of the sampled signal.

8. Sea spectra are sometimes modelled by the *Peirson-Moskowitz* spectrum, which has the form below and is usually only appropriate for deep water conditions.

$$\Gamma(\omega) = a\omega^{-5}e^{-b\omega^{-4}} \qquad 0 \le \omega \le \pi$$

   Plot the Peirson-Moskowitz spectrum in R for a few choices of parameters a and b. Compare it with the wave elevation spectra (Fig. 9.7).

10

System Identification

10.1 Purpose

Vibration is defined as an oscillatory movement of some entity about an equilibrium state. It is the means of producing sound in musical instruments, it is the principle underlying the design of loudspeakers, and it describes the response of buildings to earthquakes. The squealing of disc brakes on a car is caused by vibration. The up and down motion of a ship at sea is a low-frequency vibration. Spectral analysis provides the means for understanding and controlling vibration.

Vibration is generally caused by some external force acting on a system, and the relationship between the external force and the system response can be described by a mathematical model of the system dynamics. We can use spectral analysis to estimate the parameters of the mathematical model and then use the model to make predictions of the response of the system under different forces.

10.2 Identifying the gain of a linear system

10.2.1 Linear system

We consider systems that have clearly defined inputs and outputs, and aim to deduce the system from measurements of the inputs and outputs or to predict the output knowing the system and the input. Attempts to understand economies and to control inflation by increasing interest rates provide ambitious examples of applications of these principles.

A mathematical model of a dynamic system is linear if the output to a sum of input variables, x and y, equals the sum of the outputs corresponding to the individual inputs. More formally, a mathematical operator $\mathcal{L}$ is linear if it satisfies

$$\mathcal{L}(ax + by) = a\mathcal{L}(x) + b\mathcal{L}(y)$$

P.S.P. Cowpertwait and A.V. Metcalfe, *Introductory Time Series with R*,
Use R, DOI 10.1007/978-0-387-88698-5_10,
© Springer Science+Business Media, LLC 2009

where a and b are constants. For a linear system, the output response to a sine wave input is a sine wave of the same frequency with an amplitude that is proportional to the amplitude of the input. The ratio of the output amplitude to the input amplitude, known as the gain, and the phase lag between input and output depend on the frequency of the input, and this dependence provides a complete description of a linear system.

Many physical systems are well approximated by linear mathematical models, provided the input amplitude is not excessive. In principle, we can identify a linear model by noting the output, commonly referred to as the response, to a range of sine wave inputs. But there are practical limitations to such a procedure. In many cases, while we may be able to measure the input, we certainly cannot specify it. Examples are wave energy devices moored at sea, and the response of structures to wind forcing. Even when we can specify the input, recording the output over a range of frequencies is a slow procedure. In contrast, provided we can measure the input and output, and the input has a sufficiently broad spectrum, we can identify the linear system from spectral analysis. Also, spectral methods have been developed for non-linear systems.

A related application of spectral analysis is that we can determine the spectrum of the response if we know the system and the input spectrum. For example, we can predict the output of a wave energy device if we have a mathematical model for its dynamics and know typical sea spectra at its mooring.

10.2.2 Natural frequencies

If a system is set in motion by an initial displacement or impact, it may oscillate, and this oscillation takes place at the *natural frequency* (or frequencies) of the system. A simple example is the oscillation of a mass suspended by a spring. Linear systems have large gains at natural frequencies and, if large oscillations are undesirable, designers need to ensure that the natural frequencies of the system are far removed from forcing frequencies. Alternatively, in the case of wave energy devices, for example, the designer may aim for the natural frequencies of the device to match predominant frequencies in the sea spectrum. A common example of forcing a system at its natural frequency is pushing a child on a swing.

10.2.3 Estimator of the gain function

If a linear system is forced by a sine wave of amplitude A at frequency f, the response has an amplitude $G(f)A$, where $G(f)$ is the gain at frequency f. The ratio of the variance of the output to the variance of the input, for sine waves at this frequency, is $G(f)^2$. If the input is a stationary random process rather than a single sine wave, its variance is distributed over a range of frequencies, and this distribution is described by the spectrum. It seems intuitively reasonable to estimate the square of the gain function by the ratio of the output spectrum to the input spectrum. Consider a linear system with a single input, x_t, and a single output, y_t. The gain function can be estimated by

$$\hat{G}(f) = \sqrt{\frac{C_{yy}(f)}{C_{uu}(f)}} \tag{10.1}$$

A corollary is that the output spectrum can be estimated if the gain function is known, or has been estimated, and the input spectrum has been estimated by

$$C_{yy} = G^2 C_{uu} \tag{10.2}$$

Equation (10.2) also holds if spectra are expressed in radians rather than cycles, in which case the gain is a function $G(\omega)$ of ω.

10.3 Spectrum of an AR(p) process

Consider the deterministic part of an AR(p) model with a complex sinusoid input,

$$x_t - \alpha_1 x_{t-1} - \ldots - \alpha_p x_{t-p} = e^{i\omega t} \tag{10.3}$$

Assume a solution for x_t of the form $A e^{i\theta} e^{i\omega t}$, where A is a complex number, and substitute this into Equation (10.3) to obtain

$$A = \left(1 - \alpha_1 e^{-i\omega} - \ldots - \alpha_p e^{-i\omega p}\right)^{-1} \tag{10.4}$$

The gain function, expressed as a function of ω, is the absolute value of A. Now consider a discrete white noise input, w_t, in place of the complex sinusoid. The system is now an AR(p) process. Applying Equation (10.2), with population spectra rather than sample spectra, and noting that the spectrum of white noise with unit variance is $1/\pi$ (§9.8.1), gives

$$\Gamma_{xx}(\omega) = |A|^2 \Gamma_{ww} = \frac{1}{\pi}\left(1 - \alpha_1 e^{-i\omega} - \ldots - \alpha_p e^{-i\omega p}\right)^{-2} \qquad 0 \le \omega < \pi \tag{10.5}$$

The deterministic part of an AR(p) model is a *linear difference* equation of order p.

10.4 Simulated single mode of vibration system

The simplest linear model (SI units in parentheses) for a vibrating system is that of a mass m (N) on a spring of stiffness k (Nm^{-1}) with a damper characterised by a damping coefficient c (Nsm^{-1}). Denote the displacement of the mass by y, differentiation with respect to time by a dot placed above it, and the forcing term by x. If we apply Newton's second law of motion, equating the product of mass and acceleration with the forces acting on the mass, we obtain:

$$m\ddot{y} + c\dot{y} + ky = x \tag{10.6}$$

Equation (10.6) has the same form as that given in Exercise 1: the undamped natural frequency is $\sqrt{(k/m)}$ and the damping coefficient is $c/(2\sqrt{\{km\}})$. The gain is given by

$$G(\omega) = \left[(k - m\omega^2)^2 + c^2\omega^2\right]^{-\frac{1}{2}} \tag{10.7}$$

and the damped natural frequency, which corresponds to the maximum gain, is given by

$$\sqrt{\frac{k}{m}\left(1 - \frac{c^2}{4km}\right)} \tag{10.8}$$

These results can be derived by substituting $x = \sin(\omega t)$ and $y = G\sin(\omega t - \psi)$ into Equation (10.6) (also see Exercise 1).

Equation (10.6) represents a single mode of vibration because there is a single mass that is constrained to move in a straight line without rotating. The equation is a good model for a pendulum making small oscillations. It might be a reasonable model for vibration of a street lamp on a metal pole in gusts of wind from the same direction. It would be a poor model for vibration of a violin string because it could only describe the fundamental mode shape and would miss all of the harmonics and overtones. Nevertheless, Equation (10.6) is a widely used approximation in vibration analysis. Thomson (1993) is a nice introduction to the theory of vibration.

If we take a time step of Δ (that is, a small fraction of the unit of time) we can approximate derivatives by backward differences. Thus Equation (10.6) can be approximated by the difference equation

$$a_0 y_t + a_1 y_{t-1} + a_2 y_{t-2} = x_t \tag{10.9}$$

where

$$a_0 = \frac{m}{\Delta^2} + \frac{c}{\Delta} + k; \quad a_1 = -\frac{2m}{\Delta^2} - \frac{c}{\Delta}; \quad a_2 = \frac{m}{\Delta^2}$$

The following short R script investigates a difference equation approximation to a lightly damped system represented by Equation (10.6) with $m = 1$, $c = 1$, and $k = 16.25$. The undamped natural frequency is 4.03 and the damping coefficient is 0.124, so the damped natural frequency is 4 radians per second, assuming time is measured in seconds. The maximum gain is obtained by substituting the damped natural frequency into Equation (10.7) and is 0.250. Equation (10.6) is approximated with Equation (10.9). The input x_t is an AR(2) process with α_1 and α_2 set at 1 and -0.5, respectively, driven by Gaussian white noise with unit variance. The sampling rate is 100 per second, so the spectrum is defined from 0 to 50 Hz. The record length, n, is 100,000 and R calculates the spectrum at 50,000 points. The natural frequency is around 0.64 Hz, so the gain function is only plotted up to a frequency of 5 Hz. The arrays Freq, FreH, Omeg, and OmegH contain the discretized frequencies in cycles per sampling interval, Hz, radians per sampling interval, and radians per second, respectively. Gth is the theoretical gain of the linear system. Gemp is the empirical estimate of the gain calculated as the square root of the ratio of the output spectrum to the input spectrum. Gar is the theoretical gain of the difference equation approximation, and it is indistinguishable from the empirical estimate (Fig. 10.1). As the signals are noise-free this is not surprising. You are asked to investigate the effects of adding noise to the input and output signals in Exercise 2. The empirical estimate of the gain identifies the natural frequency accurately but slightly underestimates the maximum gain, and you are asked to investigate possible reasons for this in Exercise 3.

```
> m <- 1; c <- 1; k <- 16.25; Delta <- 0.01
> a0 <- m / Delta^2 + c / Delta + k
> a1 <- -2 * m / Delta^2 - c / Delta; a2 <- m / Delta^2
> n <- 100000
> y <- c(0, 0); x <- c(0, 0)
> set.seed(1)
> for (i in 3:n) {
    x[i] <- x[i-1] - 0.5 * x[i-2] + rnorm(1)
    y[i] <- (-a1 * y[i-1] - a2 * y[i-2]) / a0 + x[i] / a0
  }
> Sxx <- spectrum(x, span = 31)
> Syy <- spectrum(y, span = 31)
> Gemp <- sqrt( Syy$spec[1:5000] / Sxx$spec[1:5000] )
> Freq <- Syy$freq[1:5000]
> FreH <- Freq / Delta
> Omeg  <- 2 * pi * Freq
> OmegH <- 2 * pi * FreH
> Gth <- sqrt( 1/( (k-m*OmegH^2)^2 + c^2*OmegH^2 ))
> Gar <- 1 / abs( 1 + a1/a0 * exp(-Omeg*1i) + a2/a0 * exp(-Omeg*2i) )
> plot(FreH, Gth, xlab = "Frequency (Hz)", ylab = "Gain", type="l")
> lines(FreH, Gemp, lty = "dashed")
> lines(FreH, Gar, lty = "dotted")
```

10.5 Ocean-going tugboat

The motion of ships and aircraft is described by displacements along the orthogonal x, y, and z axes and rotations about these axes. The displacements are surge, sway, and heave along the x, y, and z axes, respectively. The rotations about the x, y, and z axes are roll, pitch, and yaw, respectively (Fig. 10.2). So, there are six degrees of freedom for a ship's motion in the ocean, and there are six natural frequencies. However, the natural frequencies will not usually correspond precisely to the displacements and rotations, as there is a coupling between displacements and rotations. This is typically most pronounced between heave and pitch. There will be a natural frequency with

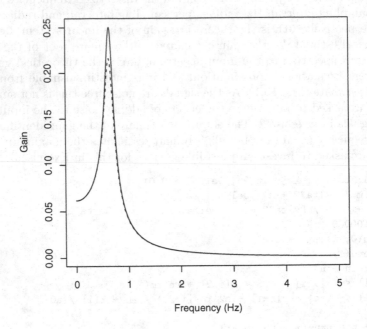

Fig. 10.1. Gain of single-mode linear system. The theoretical gain is shown by a solid line and the estimate made from the spectra obtained from the difference equation is shown by a broken line. The theoretical gain of the difference equation is plotted as a dotted line and coincides exactly with the estimate.

a corresponding mode that is predominantly heave, with a slight pitch, and another natural frequency that is predominantly pitch, with a slight heave.

Naval architects will start with computer designs and then proceed to model testing in a wave tank before building a prototype. They will have a good idea of the frequency response of the ship from the models, but this will have to be validated against sea trials. Here, we analyse some of the data from the sea trials of an ocean-going tugboat. The ship sailed over an octagonal course, and data were collected on each leg. There was an impressive array of electronic instruments and, after processing analog signals through anti-aliasing filters, data were recorded at 0.5s intervals for roll (degrees), pitch (degrees), heave (m), surge (m), sway (m), yaw (degrees), wave height (m), and wind speed (knots).

```
> www <- "http://www.massey.ac.nz/~pscowper/ts/leg4.dat"
> tug.dat <- read.table(www, header = T)
> attach(tug.dat)
> Heave.spec <- spectrum( Heave, span = sqrt( length(Heave) ),
                          log = c("no"), main = "" )
```

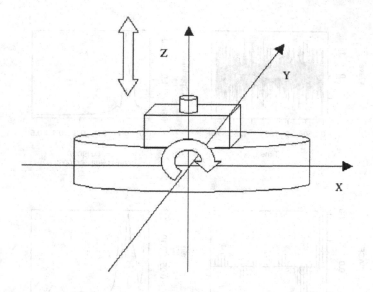

Fig. 10.2. Orthogonal axes for describing motion of a ship. Heave and pitch are shown by block arrows.

```
> Wave.spec  <- spectrum( Wave,  span = sqrt( length(Heave) ),
                                    log = c("no"), main = "" )
> G <- sqrt(Heave.spec$spec/Wave.spec$spec)
> par(mfcol = c(2, 2))
> plot( as.ts(Wave) )
> acf(Wave)
> spectrum(Wave, span = sqrt(length(Heave)), log = c("no"), main = "")
> plot(Heave.spec$freq, G, xlab="frequency Hz", ylab="Gain", type="l")
```

Figure 10.3 shows the estimated wave spectrum and the estimated gain from wave height to heave. The natural frequencies associated with the heave/pitch modes are estimated as 0.075 Hz and 0.119 Hz, and the corresponding gains from wave to heave are 0.15179 and 0.1323. In theory, the gain will approach 1 as the frequency approaches 0, but the sea spectrum has negligible components very close to 0, and no sensible estimate can be made. Also, the displacements were obtained by integrating accelerometer signals, and this is not an ideal procedure at very low frequencies.

10.6 Non-linearity

There are several reasons why the hydrodynamic response of a ship will not be precisely linear. In particular, the varying cross-section of the hull accounts

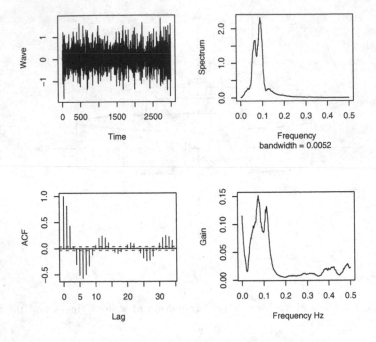

Fig. 10.3. Gain of heave from wave.

for non-linear buoyancy forces. Metcalfe et al. (2007) investigate this by fitting a regression of the heave response on lagged values of the response, squares, and cross-products of these lagged values, wave height, and wind speed. The probing method looks at the response of the fitted model to the sum of two complex sinusoids at frequencies ω_1 and ω_2. The non-linear response can be shown as a three-dimensional plot of the gain surface against frequency ω_1 and ω_2 or by a contour diagram. However, in this particular application the gain associated with the non-linear terms was small compared with the gain of the linear terms (Metcalfe et al., 2007). This is partly because the model was fitted to data taken when the ship was in typical weather conditions – under extreme conditions, when capsizing is likely, linear models are inadequate.

10.7 Exercises

1. The differential equation that describes the motion of a linear system with a single mode of vibration, such as a mass on a spring, has the general form

$$\ddot{y} + 2\zeta\Omega\dot{y} + \Omega^2 y = x$$

The parameter Ω is the undamped natural frequency, and the parameter ζ is the damping coefficient. The response is oscillatory if $\zeta < 1$.

a) Refer to Equation 10.7 and express ζ and Ω in terms of m, c, and k.

b) Suppose there is no forcing term ($x = 0$), assume that $y = e^{mt}$, and substitute into the general form of the differential equation. Show that $m = -\zeta\Omega \pm i\sqrt{[\Omega^2(1 - \zeta^2)]}$. The damped natural frequency is $\Omega\sqrt{(1 - \zeta^2)}$.

c) Take the initial condition of the unforced system as $y = 1$ when $t = 0$. Find the solution for y, and explain why this is referred to as the transient response.

d) Now consider a periodic forcing term $x = e^{i\omega t}$. Write the steady state response, y, as $y = Ae^{i(\omega t + \phi)}$. Substitute into the general form of the differential equation and show that

$$A = \left(\Omega^2 - \omega^2 + 4\zeta^2\omega^2\Omega^2\right)^{-1/2}$$

$$\tan(\Omega) = \frac{2\zeta\omega\Omega}{\Omega^2 - \omega^2}$$

2. Refer to the R script in §10.4, which compares a difference equation approximation to a model of a mass vibrating on a spring with the theoretical results. Insert another loop after that in lines 6–10 to simulate measurement noise added to the input x and y:

```
for (i in 1:n) {
  x[i] <- x[i] + nax * rnorm(1)
  y[i] <- y[i] + nay * rnorm(1)
}
```

Note that you also need to specify numerical values for the noise amplitudes, nax and nay, earlier in your script.

a) Why does the addition of noise need to be put in a separate loop?

b) How does the addition of white measurement noise to the output, but not the input, affect the estimate of the spectrum?

c) How does the addition of independent white measurement noise to both input and output affect the estimate of the spectrum?

3. The difference equation approximation used in §10.4 underestimates the maximum gain.

a) Investigate the effect of the span parameter.

b) Investigate the effect of increasing the sampling rate to 1000 per second.

c) Investigate the effect of using a centred difference approximation to the derivatives.

$$\dot{y} \approx \frac{y_{t+1} - y_{t-1}}{2\Delta}$$

11

Multivariate Models

11.1 Purpose

Data are often collected on more than one variable. For example, in economics, daily exchange rates are available for a large range of currencies, or, in hydrological studies, both rainfall and river flow measurements may be taken at a site of interest. In Chapter 10, we considered a frequency domain approach where variables are classified as inputs or outputs to some system. In this chapter, we consider time domain models that are suitable when measurements have been made on more than one time series variable. We extend the basic autoregressive model to the vector autoregressive model, which has more than one dependent time series variable, and look at methods in R for fitting such models. We consider series, called cointegrated series, that share an underlying stochastic trend, and look at suitable statistical tests for detecting cointegration. Since variables measured in time often share similar properties, regression can be used to relate the variables. However, regression models of time series variables can be misleading, so we first consider this problem in more detail before moving on to suitable models for multivariate time series.

11.2 Spurious regression

It is common practice to use regression to explore the relationship between two or more variables, and we usually seek predictor variables that either directly cause the response or provide a plausible physical explanation it. For time series variables we have to be particularly careful before ascribing any causal relationship since an apparent relationship could exist due to common extraneous factors that give rise to an underlying trend or simply because both series exhibit seasonal fluctuations. For example, the Australian electricity and chocolate production series share an increasing trend (see the following code) due to an increasing Australian population, but this does not imply that changes in one variable cause changes in the other.

P.S.P. Cowpertwait and A.V. Metcalfe, *Introductory Time Series with R*, 211
Use R, DOI 10.1007/978-0-387-88698-5_11,
© Springer Science+Business Media, LLC 2009

```
> www <- "http://www.massey.ac.nz/~pscowper/ts/cbe.dat"
> CBE <- read.table(www, header = T)
> Elec.ts <- ts(CBE[, 3], start = 1958, freq = 12)
> Choc.ts <- ts(CBE[, 1], start = 1958, freq = 12)
> plot(as.vector(aggregate(Choc.ts)), as.vector(aggregate(Elec.ts)))
> cor(aggregate(Choc.ts), aggregate(Elec.ts))
```

[1] 0.958

The high correlation of 0.96 and the scatter plot do not imply that the electricity and chocolate production variables are causally related (Fig. 11.1). Instead, it is more plausible that the increasing Australian population accounts for the increasing trend in both series. Although we can fit a regression of one variable as a linear function of the other, with added random variation, such regression models are usually termed *spurious* because of the lack of any causal relationship. In this case, it would be far better to regress the variables on the Australian population.

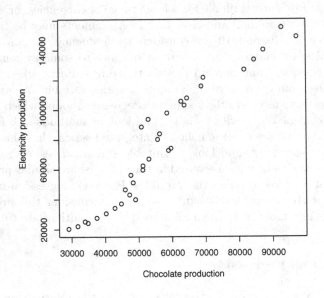

Fig. 11.1. Annual electricity and chocolate production plotted against each other.

The term spurious regression is also used when underlying stochastic trends in both series happen to be coincident, and this seems a more appropriate use of the term. Stochastic trends are a feature of an ARIMA process with a unit root (i.e., $B = 1$ is a solution of the characteristic equation). We illustrate this by simulating two independent random walks:

```
> set.seed(10); x <- rnorm(100); y <- rnorm(100)
> for(i in 2:100) {
    x[i] <- x[i-1] + rnorm(1)
    y[i] <- y[i-1] + rnorm(1) }
> plot(x, y)
> cor(x, y)
[1] 0.904
```

The code above can be repeated for different random number seeds though you will only sometimes notice spurious correlation. The seed value of 10 was selected to provide an example of a strong correlation that could have resulted by chance. The scatter plot shows how two independent time series variables might appear related when each variable is subject to stochastic trends (Fig. 11.2).

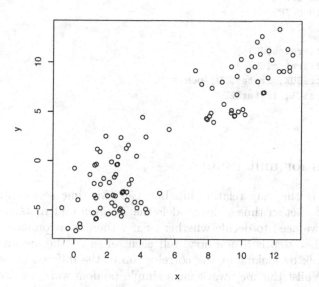

Fig. 11.2. The values of two independent simulated random walks plotted against each other. (See the code in the text.)

Stochastic trends are common in economic series, and so considerable care is required when trying to determine any relationships between the variables in multiple economic series. It may be that an underlying relationship can be justified even when the series exhibit stochastic trends because two series may be related by a common stochastic trend.

For example, the daily exchange rate series for UK pounds, the Euro, and New Zealand dollars, given for the period January 2004 to December 2007, are all per US dollar. The correlogram plots of the differenced UK and EU series indicate that both exchange rates can be well approximated by random walks (Fig. 11.3), whilst the scatter plot of the rates shows a strong linear relationship (Fig. 11.4), which is supported by a high correlation of 0.95. Since the United Kingdom is part of the European Economic Community (EEC), any change in the Euro exchange rate is likely to be apparent in the UK pound exchange rate, so there are likely to be fluctuations common to both series; in particular, the two series may share a common stochastic trend. We will discuss this phenomenon in more detail when we look at cointegration in §11.4.

```
> www <- "http://www.massey.ac.nz/~pscowper/ts/us_rates.dat"
> xrates <- read.table(www, header = T)
> xrates[1:3, ]

    UK   NZ    EU
1 0.558 1.52 0.794
2 0.553 1.49 0.789
3 0.548 1.49 0.783

> acf( diff(xrates$UK) )
> acf( diff(xrates$EU) )
> plot(xrates$UK, xrates$EU, pch = 4)
> cor(xrates$UK, xrates$EU)
[1] 0.946
```

11.3 Tests for unit roots

When investigating any relationship between two time series variables we should check whether time series models that contain unit roots are suitable. If they are, we need to decide whether or not there is a common stochastic trend. The first step is to see how well each series can be approximated as a random walk by looking at the correlogram of the differenced series (e.g., Fig. 11.3). Whilst this may work for a simple random walk, we have seen in Chapter 7 that stochastic trends are a feature of any time series model with a unit root $B = 1$ as a solution of the characteristic equation, which would include more complex ARIMA processes.

Dickey and Fuller developed a test of the null hypothesis that $\alpha = 1$ against an alternative hypothesis that $\alpha < 1$ for the model $x_t = \alpha x_{t-1} + u_t$ in which u_t is white noise. A more general test, which is known as the augmented Dickey-Fuller test (Said and Dickey, 1984), allows the differenced series u_t to be any stationary process, rather than white noise, and approximates the stationary process with an AR model. The method is implemented in R by the function adf.test within the tseries library. The null hypothesis of a unit root cannot be rejected for our simulated random walk x:

```
> library(tseries)
> adf.test(x)
```

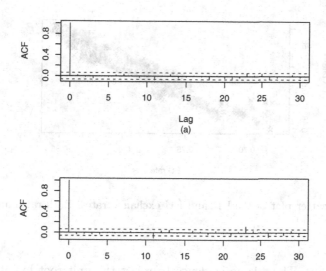

Fig. 11.3. Correlograms of the differenced exchange rate series: (a) UK rate; (b) EU rate.

```
Augmented Dickey-Fuller Test

data:  x
Dickey-Fuller = -2.23, Lag order = 4, p-value = 0.4796
alternative hypothesis: stationary
```

This result is not surprising since we would only expect 5% of simulated random walks to provide evidence against a null hypothesis of a unit root at the 5% level. However, when we analyse physical time series rather than realisations from a known model, we should never mistake lack of evidence against a hypothesis for a demonstration that the hypothesis is true. The test result should be interpreted with careful consideration of the length of the time series, which determines the power of the test, and the general context. The null hypothesis of a unit root is favoured by economists because many financial time series are better approximated by random walks than by a stationary process, at least in the short term.

An alternative to the augmented Dickey-Fuller test, known as the Phillips-Perron test (Perron, 1988), is implemented in the R function `pp.test`. The distinction between the two tests is that the Phillips-Perron procedure estimates the autocorrelations in the stationary process u_t directly (using a kernel smoother) rather than assuming an AR approximation, and for this reason the Phillips-Perron test is described as semi-parametric. Critical values of the test statistic are either based on asymptotic theory or calculated from exten-

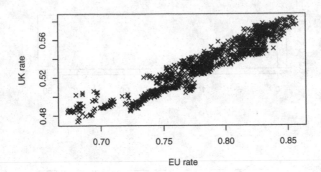

Fig. 11.4. Scatter plot of the UK and EU exchange rates. Both rates are per US dollar.

sive simulations. There is no evidence to reject the unit root hypothesis, so we conclude that the UK pound and Euro exchange rates are both likely to contain unit roots.

```
> pp.test(xrates$UK)

        Phillips-Perron Unit Root Test

data:  xrates$UK
Dickey-Fuller Z(alpha) = -10.6, Truncation lag parameter = 7,
p-value = 0.521
alternative hypothesis: stationary

> pp.test(xrates$EU)

        Phillips-Perron Unit Root Test

data:  xrates$EU
Dickey-Fuller Z(alpha) = -6.81, Truncation lag parameter = 7,
p-value = 0.7297
alternative hypothesis: stationary
```

11.4 Cointegration

11.4.1 Definition

Many multiple time series are highly correlated in time. For example, in §11.2 we found the UK pound and Euro exchange rates very highly correlated. This is explained by the similarity of the two economies relative to the US economy. Another example is the high correlation between the Australian electricity and chocolate production series, which can be reasonably attributed to an increasing Australian population rather than a causal relationship. In addition, we demonstrated that two series that are independent and contain unit roots (e.g., they follow independent random walks) can show an apparent linear relationship, due to chance similarity of the random walks over the period of the time series, and stated that such a correlation would be spurious. However, as demonstrated by the analysis of the UK pounds and Euro exchange rates, it is quite possible for two series to contain unit roots and be related. Such series are said to be *cointegrated*. In the case of the exchange rates, a stochastic trend in the US economy during a period when the European economy is relatively stable will impart a common, complementary, stochastic trend to the UK pound and Euro exchange rates. We now state the precise definition of cointegration.

Two non-stationary time series $\{x_t\}$ and $\{y_t\}$ are cointegrated if some linear combination $ax_t + by_t$, with a and b constant, is a stationary series.

As an example consider a random walk $\{\mu_t\}$ given by $\mu_t = \mu_{t-1} + w_t$, where $\{w_t\}$ is white noise with zero mean, and two series $\{x_t\}$ and $\{y_t\}$ given by $x_t = \mu_t + w_{x,t}$ and $y_t = \mu_t + w_{y,t}$, where $\{w_{x,t}\}$ and $\{w_{y,t}\}$ are independent white noise series with zero mean. Both series are non-stationary, but their difference $\{x_t - y_t\}$ is stationary since it is a finite linear combination of independent white noise terms. Thus the linear combination of $\{x_t\}$ and $\{y_t\}$, with $a = 1$ and $b = -1$, produced a stationary series, $\{w_{x,t} - w_{y,t}\}$. Hence $\{x_t\}$ and $\{y_t\}$ are cointegrated and share the underlying stochastic trend $\{\mu_t\}$.

In R, two series can be tested for cointegration using the Phillips-Ouliaris test implemented in the function po.test within the tseries library. The function requires the series be given in matrix form and produces the results for a test of the null hypothesis that the two series are not cointegrated. As an example, we simulate two cointegrated series x and y that share the stochastic trend mu and test for cointegration using po.test:

```
> x <- y <- mu <- rep(0, 1000)
> for (i in 2:1000) mu[i] <- mu[i - 1] + rnorm(1)
> x <- mu + rnorm(1000)
> y <- mu + rnorm(1000)
> adf.test(x)$p.value

[1] 0.502

> adf.test(y)$p.value

[1] 0.544

> po.test(cbind(x, y))

        Phillips-Ouliaris Cointegration Test
```

```
data:  cbind(x, y)
Phillips-Ouliaris demeaned = -1020, Truncation lag parameter = 9,
p-value = 0.01
```

In the example above, the conclusion of the adf.test is to retain the null hypothesis that the series have unit roots. The po.test provides evidence that the series are cointegrated since the null hypothesis is rejected at the 1% level.

11.4.2 Exchange rate series

The code below is an analysis of the UK pound and Euro exchange rate series. The Phillips-Ouliaris test shows there is evidence that the series are cointegrated, which justifies the use of a regression model. An ARIMA model is then fitted to the residuals of the regression model. The ar function is used to determine the best order of an AR process. We can investigate the adequacy of our cointegrated model by using R to fit a more general ARIMA process to the residuals. The best-fitting ARIMA model has $d = 0$, which is consistent with the residuals being a realisation of a stationary process and hence the series being cointegrated.

```
> po.test(cbind(xrates$UK, xrates$EU))

        Phillips-Ouliaris Cointegration Test

data:  cbind(xrates$UK, xrates$EU)
Phillips-Ouliaris demeaned = -21.7, Truncation lag parameter = 10,
p-value = 0.04118

> ukeu.lm <- lm(xrates$UK ~ xrates$EU)
> ukeu.res <- resid(ukeu.lm)
> ukeu.res.ar <- ar(ukeu.res)
> ukeu.res.ar$order

[1] 3

> AIC(arima(ukeu.res, order = c(3, 0, 0)))

[1] -9886

> AIC(arima(ukeu.res, order = c(2, 0, 0)))

[1] -9886

> AIC(arima(ukeu.res, order = c(1, 0, 0)))

[1] -9880

> AIC(arima(ukeu.res, order = c(1, 1, 0)))

[1] -9876
```

Comparing the AICs for the AR(2) and AR(3) models, it is clear there is little difference and that the AR(2) model would be satisfactory. The example above also shows that the AR models provide a better fit to the residual series than the ARIMA(1, 1, 0) model, so the residual series may be treated as stationary. This supports the result of the Phillips-Ouliaris test since a linear combination of the two exchange rates, obtained from the regression model, has produced a residual series that appears to be a realisation of a stationary process.

11.5 Bivariate and multivariate white noise

Two series $\{w_{x,t}\}$ and $\{w_{y,t}\}$ are *bivariate* white noise if they are stationary and their cross-covariance $\gamma_{xy}(k) = \text{Cov}(w_{x,t}, w_{y,t+k})$ satisfies

$$\gamma_{xx}(k) = \gamma_{yy}(k) = \gamma_{xy}(k) = 0 \qquad \text{for all } k \neq 0 \tag{11.1}$$

In the equation above, $\gamma_{xx}(0) = \gamma_{yy}(0) = 1$ and $\gamma_{xy}(0)$ may be zero or non-zero. Hence, bivariate white noise series $\{w_{x,t}\}$ and $\{w_{y,t}\}$ may be regarded as white noise when considered individually but when considered as a pair may be cross-correlated at lag 0.

The definition of bivariate white noise readily extends to *multivariate* white noise. Let $\gamma_{ij}(k) = \text{Cov}(w_{i,t}, w_{j,t+k})$ be the cross-correlation between the series $\{w_{i,t}\}$ and $\{w_{j,t}\}$ $(i, j = 1, \ldots n)$. Then stationary series $\{w_{1,t}\}$, $\{w_{2,t}\}$, ..., $\{w_{n,t}\}$ are multivariate white noise if each individual series is white noise and, for each pair of series $(i \neq j)$, $\gamma_{ij}(k) = 0$ for all $k \neq 0$. In other words, multivariate white noise is a sequence of independent draws from some multivariate distribution.

Multivariate Gaussian white noise can be simulated with the `rmvnorm` function in the `mvtnorm` library. The function may take a mean and covariance matrix as a parameter input, and the dimensions of these determine the dimension of the output matrix. In the following example, the covariance matrix is 2×2, so the output variable `x` is bivariate with 1000 simulated white noise values in each of two columns. An arbitrary value of 0.8 is chosen for the correlation to illustrate the use of the function.

```
> library(mvtnorm)
> cov.mat <- matrix(c(1, 0.8, 0.8, 1), nr = 2)
> w <- rmvnorm(1000, sigma = cov.mat)
> cov(w)

        [,1]  [,2]
[1,] 1.073 0.862
[2,] 0.862 1.057

> wx <- w[, 1]
> wy <- w[, 2]
> ccf(wx, wy, main = "")
```

The ccf function verifies that the cross-correlations are approximately zero for all non-zero lags (Fig. 11.5). As an exercise, check that the series in each column of x are approximately white noise using the acf function.

One simple use of bivariate or multivariate white noise is in the method of *prewhitening*. Separate SARIMA models are fitted to multiple time series variables so that the residuals of the fitted models appear to be a realisation of multivariate white noise. The SARIMA models can then be used to forecast the expected values of each time series variable, and multivariate simulations can be produced by adding multivariate white noise terms to the forecasts. The method works well provided the multiple time series have no common stochastic trends and the cross-correlation structure is restricted to the error process.

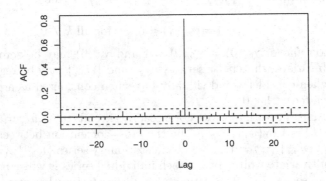

Fig. 11.5. Cross-correlation of simulated bivariate Gaussian white noise

11.6 Vector autoregressive models

Two time series, $\{x_t\}$ and $\{y_t\}$, follow a vector autoregressive process of order 1 (denoted VAR(1)) if

$$
\begin{aligned}
x_t &= \theta_{11} x_{t-1} + \theta_{12} y_{t-1} + w_{x,t} \\
y_t &= \theta_{21} x_{t-1} + \theta_{22} y_{t-1} + w_{y,t}
\end{aligned}
\tag{11.2}
$$

where $\{w_{x,t}\}$ and $\{w_{y,t}\}$ are bivariate white noise and θ_{ij} are model parameters. If the white noise sequences are defined with mean 0 and the process is stationary, both time series $\{x_t\}$ and $\{y_t\}$ have mean 0 (Exercise 1). The simplest way of incorporating a mean is to define $\{x_t\}$ and $\{y_t\}$ as deviations from mean values. Equation (11.2) can be rewritten in matrix notation as

$$
\mathbf{Z}_t = \boldsymbol{\Theta} \mathbf{Z}_{t-1} + \mathbf{w}_t
\tag{11.3}
$$

where

$$\mathbf{Z}_t = \begin{pmatrix} x_t \\ y_t \end{pmatrix} \qquad \boldsymbol{\Theta} = \begin{pmatrix} \theta_{11} & \theta_{12} \\ \theta_{21} & \theta_{22} \end{pmatrix} \qquad \mathbf{w}_t = \begin{pmatrix} w_{x,t} \\ w_{y,t} \end{pmatrix}$$

Equation (11.3) is a vector expression for an AR(1) process; i.e., the process is vector autoregressive. Using the backward shift operator, Equation (11.3) can also be written

$$(\mathbf{I} - \boldsymbol{\Theta}B)\mathbf{Z}_t = \boldsymbol{\theta}(B)\mathbf{Z}_t = \mathbf{w}_t \qquad (11.4)$$

where $\boldsymbol{\theta}$ is a matrix polynomial of order 1 and $\mathbf{I}$ is the 2×2 identity matrix. A VAR(1) process can be extended to a VAR(p) process by allowing $\boldsymbol{\theta}$ to be a matrix polynomial of order p. A VAR(p) model for m time series is also defined by Equation (11.4), in which $\mathbf{I}$ is the $m \times m$ identity matrix, $\boldsymbol{\theta}$ is a polynomial of $m \times m$ matrices of parameters, $\mathbf{Z}_t$ is an $m \times 1$ matrix of time series variables, and w_t is multivariate white noise. For a VAR model, the characteristic equation is given by a determinant of a matrix. Analogous to AR models, a VAR(p) model is stationary if the roots of the determinant $|\boldsymbol{\theta}(x)|$ all exceed unity in absolute value. For the VAR(1) model, the determinant is given by

$$\begin{vmatrix} 1 - \theta_{11}x & -\theta_{12}x \\ -\theta_{21}x & 1 - \theta_{22}x \end{vmatrix} = (1 - \theta_{11}x)(1 - \theta_{22}x) - \theta_{12}\theta_{21}x^2 \qquad (11.5)$$

The R functions polyroot and Mod can be used to test whether a VAR model is stationary, where the function polyroot just takes a vector of polynomial coefficients as an input parameter. For example, consider the VAR(1) model with parameter matrix $\boldsymbol{\Theta} = \begin{pmatrix} 0.4 & 0.3 \\ 0.2 & 0.1 \end{pmatrix}$. Then the characteristic equation is given by

$$\begin{vmatrix} 1 - 0.4x & -0.3x \\ -0.2x & 1 - 0.1x \end{vmatrix} = 1 - 0.5x - 0.02x^2 \qquad (11.6)$$

The absolute value of the roots of the equation is given by

```
> Mod(polyroot(c(1, -0.5, -0.02)))
```

```
[1]  1.86 26.86
```

From this we can deduce that the VAR(1) model is stationary since both roots exceed unity in absolute value.

The parameters of a VAR(p) model can be estimated using the ar function in R, which selects a best-fitting order p based on the smallest AIC. Using the simulated bivariate white noise process of §11.5 and the parameters from the stationary VAR(1) model given above, a VAR(1) process is simulated below and the parameters from the simulated series estimated using ar.

```
> x <- y <- rep(0, 1000)
> x[1] <- wx[1]
> y[1] <- wy[1]
> for (i in 2:1000) {
```

```
        x[i] <- 0.4 * x[i - 1] + 0.3 * y[i - 1] + wx[i]
        y[i] <- 0.2 * x[i - 1] + 0.1 * y[i - 1] + wy[i]
   }
> xy.ar <- ar(cbind(x, y))
> xy.ar$ar[, , ]


      x     y
x 0.399 0.321
y 0.208 0.104
```

As expected, the parameter estimates are close to the underlying model values. If the simulation is repeated many times with different realisations of the bivariate white noise, the sampling distribution of the estimators of the parameters in the model can be approximated by the histograms of the estimates together with the correlations between estimates. This is the principle used to construct bootstrap confidence intervals for model parameters when they have been estimated from time series.

The bootstrap simulation is set up using point estimates of the parameters in the model, including the variance of the white noise terms. Then time series of the same length as the historical records are simulated and the parameters estimated. A $(1 - \alpha) \times 100\%$ confidence interval for a parameter is between the lower and upper $\alpha/2$ quantiles of the empirical sampling distribution of its estimates.

11.6.1 VAR model fitted to US economic series

A quarterly US economic series (1954–1987) is available within the tseries library. A best-fitting VAR model is fitted to the (mean-adjusted) gross national product (GNP) and real money (M1) in the following example.[1] Ordinary least squares is used to fit the model to the mean adjusted series – with dmean set to TRUE and intercept set to FALSE since the latter parameter will not be required.

```
> library(tseries)
> data(USeconomic)
> US.ar <- ar(cbind(GNP, M1), method="ols", dmean=T, intercept=F)
> US.ar$ar

, , GNP


        GNP       M1
1   1.27181  -0.0338
2  -0.00423   0.0635
3  -0.26715  -0.0286


, , M1
```

---

[1] *Real money* means income adjusted by inflation.

```
      GNP      M1
1   1.167   1.588
2  -0.694  -0.484
3  -0.510  -0.129

> acf(US.ar$res[-c(1:3), 1])
> acf(US.ar$res[-c(1:3), 2])
```

From the code above, we see that the best-fitting VAR model is of order 3. The correlogram of the residual series indicates that the residuals are approximately bivariate white noise, thus validating the assumptions for a VAR model (Fig. 11.6).

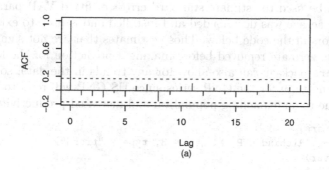

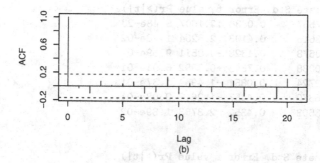

Fig. 11.6. Residual correlograms for the VAR(3) model fitted to the US economic series: (a) residuals for GNP; (b) residuals for M1.

To check for stationarity, the characteristic function can be evaluated using the determinant:

$$\left| \begin{pmatrix} 1 & 0 \\ 0 & 1 \end{pmatrix} - \begin{pmatrix} 1.272 & -0.03383 \\ 1.167 & 1.588 \end{pmatrix} x - \begin{pmatrix} -0.004230 & 0.06354 \\ -0.6942 & -0.4839 \end{pmatrix} x^2 \right|$$

$$-\begin{pmatrix} -0.2672 & -0.02859 \\ -0.5103 & -0.1295 \end{pmatrix} x^3 \Bigg|$$

$$= 1 - 2.859x + 2.547x^2 - 0.3232x^3 - 0.5265x^4 + 0.1424x^5 + 0.01999x^6$$

From this it can be verified that the fitted VAR(3) model is stationary since all the roots exceed unity in absolute value:

```
> Mod( polyroot(c(1,-2.859,2.547,-0.3232, -0.5265, 0.1424, 0.01999)) )
[1] 1.025269 1.025269 1.257038 1.598381 2.482308 9.541736
```

At the time of writing, an algorithm was not available for extracting standard errors of VAR parameter estimates from an `ar` object. Estimates of these errors could be obtained using a bootstrap method or a function from another library. In the `vars` package (Pfaff, 2008), available on the R website, the `VAR` function can be used to estimate standard errors of fitted VAR parameters. Hence, this package was downloaded and installed and is used to extract the standard errors in the code below. Those estimates that are not significantly different from zero are removed before making a prediction for the following year. The `vars` package can also allow for any trends in the data, so we also include a trend term for the GNP series since US GNP will tend to increase with time due to an expanding population and increased productivity.

```
> library(vars)
> US.var <- VAR(cbind(GNP, M1), p = 3, type = "trend")
> coef(US.var)

$GNP
        Estimate Std. Error t value Pr(>|t|)
GNP.l1  1.07537    0.0884 12.1607 5.48e-23
M1.l1   1.03615    0.4103  2.5254 1.28e-02
GNP.l2 -0.00678    0.1328 -0.0511 9.59e-01
M1.l2  -0.30038    0.7543 -0.3982 6.91e-01
GNP.l3 -0.12724    0.0851 -1.4954 1.37e-01
M1.l3  -0.56370    0.4457 -1.2648 2.08e-01
trend   1.03503    0.4352  2.3783 1.89e-02

$M1
        Estimate Std. Error t value Pr(>|t|)
GNP.l1  -0.0439    0.0191 -2.298 2.32e-02
M1.l1    1.5923    0.0887 17.961 1.51e-36
GNP.l2   0.0616    0.0287  2.148 3.36e-02
M1.l2   -0.4891    0.1630 -3.001 3.25e-03
GNP.l3  -0.0175    0.0184 -0.954 3.42e-01
M1.l3   -0.1041    0.0963 -1.081 2.82e-01
trend    0.0116    0.0940  0.123 9.02e-01

> US.var <- VAR(cbind(GNP, M1), p = 2, type = "trend")
> coef(US.var)
```

```
$GNP
         Estimate Std. Error t value Pr(>|t|)
GNP.l1     1.141      0.0845   13.51 1.83e-26
M1.l1      1.330      0.3391    3.92 1.41e-04
GNP.l2    -0.200      0.0823   -2.43 1.67e-02
M1.l2     -1.157      0.3488   -3.32 1.19e-03
trend      1.032      0.4230    2.44 1.61e-02

$M1
         Estimate Std. Error t value Pr(>|t|)
GNP.l1 -0.03372      0.0181 -1.8623 6.48e-02
M1.l1   1.64898      0.0727 22.6877 7.33e-47
GNP.l2  0.03419      0.0176  1.9384 5.48e-02
M1.l2  -0.65016      0.0748 -8.6978 1.35e-14
trend   0.00654      0.0906  0.0722 9.43e-01

> acf(resid(US.var)[, 1])
> acf(resid(US.var)[, 2])
```

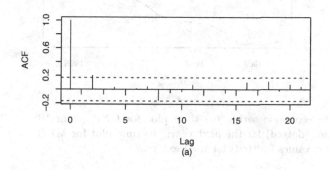
(a)

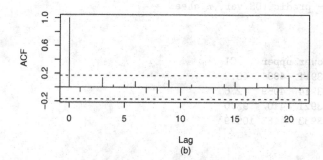
(b)

Fig. 11.7. Residual correlograms for the VAR(2) model fitted to the US economic series: (a) residuals for GNP; (b) residuals for M1.

Below we give the predicted values for the next year of the series, which are then added to a time series plot for each variable (Fig. 11.8).

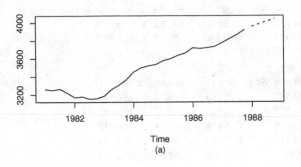

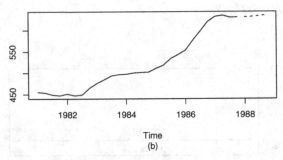

Fig. 11.8. US economic series: (a) time plot for GNP (from 1981) with added predicted values (dotted) for the next year; (b) time plot for M1 (from 1981) with added predicted values (dotted) for the next year.

```
> US.pred <- predict(US.var, n.ahead = 4)
> US.pred

$GNP
     fcst lower upper    CI
[1,] 3958  3911  4004  46.2
[2,] 3986  3914  4059  72.6
[3,] 4014  3921  4107  93.0
[4,] 4043  3933  4153 109.9

$M1
     fcst lower upper   CI
[1,]  631   621   641  9.9
[2,]  632   613   651 19.0
[3,]  634   606   661 27.5
[4,]  636   601   671 35.1
```

```
> GNP.pred <- ts(US.pred$fcst$GNP[, 1], st = 1988, fr = 4)
> M1.pred <- ts(US.pred$fcst$M1[, 1], st = 1988, fr = 4)
> ts.plot(cbind(window(GNP, start = 1981), GNP.pred), lty = 1:2)
> ts.plot(cbind(window(M1, start = 1981), M1.pred), lty = 1:2)
```

11.7 Summary of R commands

adf.test	Dickey-Fuller test for unit roots
pp.test	Phillips-Perron test for unit roots
rmvnorm	multivariate white noise simulation
po.test	Phillips-Ouliaris cointegration test
ar	Fits the VAR model based on the smallest AIC
VAR	Fits the VAR model based on least squares (vars package required)

11.8 Exercises

1. Show that if a VAR(1) process driven by white noise with mean 0, as defined in Equation 11.5, is stationary, then it has a mean of 0. Deduce that if a VAR(p) process driven by white noise with mean 0 is stationary, then it has a mean of 0. [Hint: Take expected values of both sides of Equation 11.5 and explain why the inverse of $\mathbf{I} - \boldsymbol{\Theta}$ exists.]

2. For what values of a is the model below stationary?

$$x_t = 0.9x_{t-1} + ay_{t-1} + w_{x,t}$$
$$y_t = ax_{t-1} + 0.9y_{t-1} + w_{y,t}$$

3. This question uses the data in stockmarket.dat, which contains stock market data for seven cities for the period January 6, 1986 to December 31, 1997. Download the data via the book website and put the data into a variable in R.

   a) Use an appropriate statistical test to test whether the London and/or the New York series have unit roots. Does the evidence from the statistical tests suggest the series are stationary or non-stationary?

   b) Let $\{x_t\}$ represent the London series (Lond) and $\{y_t\}$ the New York series (NY). Fit the following VAR(1) model, giving a summary output containing the fitted parameters and any appropriate statistical tests:

$$x_t = a_0 + a_1 x_{t-1} + a_2 y_{t-1} + w_{x,t}$$
$$y_t = b_0 + b_1 x_{t-1} + b_2 y_{t-1} + w_{y,t}$$

c) Which series influences the other the most? Why might this happen?
d) Test the London and New York series for cointegration.
e) Fit the model below, giving a summary of the model parameters and any appropriate statistical tests.

$$x_t = a_0 + a_1 y_t + w_t$$

f) Test the residual series for the previous fitted model for unit roots. Does this support or contradict the result in part (d)? Explain your answer.

4. a) Using the `VAR` function in the `vars` package, fit a multivariate VAR model to the four economic variables in the Canadian data (which can be loaded from within the `vars` package with the command `data(Canada)`).
   b) Using the fitted VAR model, make predictions for the next year. Add these predictions to a time series plot of each variable.

5. a) Fit an ARIMA$(1, 1, 0)(1, 1, 1)_{12}$ model to the logarithm of the electricity production series. Verify that the residuals are approximately white noise.
   b) Fit the same model as in (a) to the logarithm of the chocolate production series. Again, verify that the residuals are approximately white noise.
   c) Plot the cross-correlogram of the residuals of the two fitted ARIMA models, and verify that the lag 0 correlation is significantly different from zero. Give a possible reason why this may happen.
   d) Forecast values for the next month for each series, and add a simulated bivariate white noise term to each forecast. This gives one possible realisation. Repeat the process ten times to give ten possible future scenarios for the next month's production for each series.

12

State Space Models

12.1 Purpose

The state space formulation for time series models is quite general and encompasses most of the models we have considered so far. However, it is usually simpler to use the specific time series models we have already introduced when they are appropriate for the physical situation. Here, we shall focus on applications for which we require parameters to adapt over time, and to do so more quickly than in a Holt-Winters model. The recent turmoil on the world's stock exchanges[1] is a dramatic reminder that time series are subject to sudden changes. Another desirable feature of state space models is that they can incorporate time series of predictor variables in a straightforward manner.

Control engineers have used a state space representation of physical systems as input, state, and output variables related by first-order linear differential equations since the 1950s, and Kalman and Bucy published their famous paper on filtering in 1961 (Kalman and Bucy, 1961). Plackett (1950) published related, but less general, work on the adaptive estimation of coefficients in regression models and gave some historical background to the problem. In the control context, the state variables define the dynamics of some physical system and might, for instance, be displacements and velocities. Typically, only some of these state variables can be measured directly, and these measurements are subject to noise. The objective of the Kalman filter is to infer values for all the state variables from the noisy measurements. The estimated values of the state variables are then used to control the system. Feedback control systems are the essence of robotics, and some applications are cruise control in automobiles, autopilots in aircraft, and the planetary explorer Rover Sojourner – the Mars Pathfinder Mission was launched on the December 4, 1996.

---

[1] Notable financial events in 2008 included the US government takeover of Fannie Mae and Freddie Mac on September 7, the rejection of the first bailout bill by the US House of Representatives on September 29, and the passing of the US Emergency Economic Stabilization Act of 2008 on October 3.

P.S.P. Cowpertwait and A.V. Metcalfe, *Introductory Time Series with R*,
Use R, DOI 10.1007/978-0-387-88698-5_12,
© Springer Science+Business Media, LLC 2009

The chemical process industry provides many other applications for control engineers. Typically, states will be concentrations, temperatures, and pressures, and the controller will actuate burners, stirrers, and pumps. Digital computers are an essential feature of modern control systems, and discrete-time models tend to be used in place of continuous-time models, with differences replacing derivatives and time series replacing continuous (analog) signals.

In this chapter, we focus on economic time series. Usually, the states will be unknown coefficients in the linear models and the equations that represent changes in states will be rather simple. Nevertheless, the concept of such parameters changing rather than being fixed is a departure from most of the models we have considered so far, the exception being the Holt-Winters forecasting method. A Bayesian approach is ideal for the development of a state space model.

12.2 Linear state space models

12.2.1 Dynamic linear model

We adopt the notation used in Pole et al. (1994), who refer to state space models as dynamic linear models. The values of the state at time t are represented by a column matrix θ_t and are a linear combination of the values of the state at time $t-1$ and random variation (system noise) from a multivariate normal distribution. The linear combination of values of the state at time $t-1$ is defined by G_t, and the variance-covariance matrix of the system noise is W_t. The observation at time t is denoted by a column matrix y_t that is a linear combination of the states, determined by a matrix F_t, and random variation (measurement noise) from a normal distribution with variance-covariance matrix V_t. The random variation has mean zero and is uncorrelated over time. All the matrices can be time varying, but in many applications G_t is constant. The state space model is summarised by the equations

$$y_t = F_t'\theta_t + v_t$$
$$\theta_t = G_t\theta_{t-1} + w_t \tag{12.1}$$

where $\theta_0 \sim N(m_0, C_0)$, $v_t \sim N(0, V_t)$, and $w_t \sim N(0, W_t)$.

A specific, but very useful, application of state space models is to generalise regression models so that the parameters can vary over time. For example, the sales manager in a house building company might use the following model to allow for the influence of a general level (L) of sales in the sector and the company's own pricing (P) policy on the company's house sales (S):

$$S_t = L_t + \beta_t P_t + v_t$$
$$L_t = L_{t-1} + \Delta L_t$$
$$\beta_t = \beta_{t-1} + \Delta\beta_t \tag{12.2}$$

The first equation is a linear regression with price as the predictor variable. However, the model allows the intercept term, the level, and the coefficient of price to vary over time, and this makes it far more realistic for the house building market. The v_t, ΔL_t, and $\Delta \beta_t$ are random deviations with mean zero that are independent over time, although ΔL_t and $\Delta \beta_t$ can be correlated. The relative magnitudes of the variances of these components of error, which are the entries in the matrices V_t and W_t, determine the variability of the parameters. If $W_t = 0$, the state space model reduces to the standard regression with constant parameters. In state space form

$$y_t = S_t \qquad \theta_t = \begin{pmatrix} L_t \\ \beta_t \end{pmatrix} \qquad w_t = \begin{pmatrix} \Delta L_t \\ \Delta \beta_t \end{pmatrix} \qquad F_t = \begin{pmatrix} 1 \\ P_t \end{pmatrix} \qquad G = \begin{pmatrix} 1 & 0 \\ 0 & 1 \end{pmatrix}$$

The subscript t on the matrix G is redundant, as G is constant in this application.

The system is said to be observable if it is possible to infer the values of all the components of the state from the noisy observations (Exercise 5). If the system is observable, we can distinguish prediction, filtering, and smoothing. Prediction is the forecasting of future values of the state, filtering is making the best estimate of the current values of the state from the record of observations, including the current observation, and smoothing is making the best estimates of past values of the state given the record of observations. Filtering is particularly important because it is the basis for control algorithms and forecasting.

12.2.2 Filtering*

Let D_t represent the data up until time t. In most applications, the data are the time series of observations, but the notation does allow for the time series to be augmented by any additional information. In the following, we express the data up until time t, D_t, as the combination of data up until time $t - 1$ and the observation at time t, (D_{t-1}, y_t). Bayes's Theorem gives

$$p(\theta_t \mid D_{t-1}, y_t) = \frac{p(y_t \mid \theta_t)\, p(\theta_t \mid D_{t-1})}{p(y_t)} \qquad (12.3)$$

and is usually applied without the normalising constant, so that we can write

$$p(\theta_t \mid D_{t-1}, y_t) \propto p(y_t \mid \theta_t)\, p(\theta_t \mid D_{t-1}) \qquad (12.4)$$

That is, the posterior density of the state at time t, given data up until time t, is proportional to the product of the probability density of the observation at time t given the state at time t, referred to as the likelihood, and the prior density of the state at time t, given data up until time $t-1$. If the prior distribution and the likelihood are both normal, Bayes's Theorem provides a nice analytic form. In the univariate case, the mean of the posterior distribution is a weighted mean of the mean of the prior distribution and the observation with weights proportional to their precisions.[2] Also, the precision of the posterior distribution is the sum of the precision of the prior distribution and the precision of the observation (Exercise 6). The extension of this result to the multivariate normal distribution leads to the result

$$\theta_t \mid D_t \sim N(m_t, C_t) \tag{12.5}$$

where m_t and C_t are calculated iteratively, for t from 1 up to n, from the following algorithm, which is known as the Kalman filter. Remember that m_0 and C_0 are specified as part of the model.

Kalman filter

The prior distribution for θ_t, the likelihood, and the posterior distribution for θ_t are (respectively)

$$\theta_t \mid D_{t-1} \sim N(a_t, R_t) \qquad y_t \mid \theta_t \sim N(F_t'\theta_t, V_t) \qquad \theta_t \mid D_t \sim N(m_t, C_t)$$

Then, for $t = 1, \ldots$, the algorithm is given by

$$
\begin{aligned}
a_t &= G_t m_{t-1} & f_t &= F_t' a_t \\
R_t &= G_t C_{t-1} G_t' + W_t & Q_t &= F_t' R_t F_t + V_t \\
e_t &= y_t - f_t & A_t &= R_t F_t Q_t^{-1} \\
m_t &= a_t + A_t e_t & C_t &= R_t - A_t Q_t A_t'
\end{aligned}
$$

In this algorithm, f_t is the forecast value of the observation at time t, the forecast being made at time $t - 1$. It follows that e_t is the forecast error. The posterior mean is a weighted sum of the prior mean and the forecast error. Notice that the variance of the posterior distribution, C_t, is less than the variance of the prior distribution, R_t.

12.2.3 Prediction*

Predictions start from the posterior estimate of the state, obtained from the Kalman filter, on the day (t) on which the forecast is made. We can make one-step-ahead forecasts in the following manner:

$$E[y_{t+1} \mid D_t] = E\left[F_{t+1}'\theta_{t+1} + v_{t+1} \mid D_t\right] = F_{t+1}'E[\theta_{t+1} \mid D_t] = F_{t+1}'a_{t+1}$$
$$= f_{t+1} \tag{12.6}$$

$$\operatorname{Var}[y_{t+1} \mid D_t] = \operatorname{Var}\left[F_{t+1}'\theta_{t+1} + v_{t+1} \mid D_t\right] = F_{t+1}'\operatorname{Var}[\theta_{t+1} \mid D_t]F_{t+1} + V_{t+1}$$
$$= F_{t+1}'R_{t+1}F_{t+1} + V_{t+1}$$
$$= Q_{t+1} \tag{12.7}$$

---

[2] Precision is the reciprocal of the variance.

The general formula for a forecast at time t for k steps ahead is

$$y_{t+k} \mid D_t \sim N(f_{t+k \mid t}, Q_{t+k \mid t}) \tag{12.8}$$

where

$$
\begin{aligned}
f_{t+k \mid t} &= F'_{t+k} G^{k-1} a_{t+1} \\
Q_{t+k \mid t} &= F'_{t+k} R_{t+k \mid t} F_{t+k} + V_{t+k} \\
R_{t+k \mid t} &= G^{k-1} R_{t+1} (G^{k-1})' + \sum_{j=2}^{k} G^{k-j} W_{t+j} (G^{k-j})'
\end{aligned} \tag{12.9}
$$

12.2.4 Smoothing*

The optimal smoothing algorithm follows from a nice application of elementary probability. We demonstrate this for one step back, and the general case proceeds in the same way. To begin with, we use the rule of total probability

$$p(\theta_{t-1} \mid D_t) = \int p(\theta_{t-1} \mid \theta_t, D_t) \, p(\theta_t \mid D_t) \, d\theta_t \tag{12.10}$$

The $p(\theta_t \mid D_t)$ in the integrand on the right-hand side is available from the Kalman filter, so we only need to consider further

$$p(\theta_{t-1} \mid \theta_t, D_t) = p(\theta_{t-1} \mid \theta_t, D_{t-1}) \tag{12.11}$$

because y_t provides no further information once θ_t is known. Now we can apply Bayes's Theorem:

$$p(\theta_{t-1} \mid \theta_t, D_{t-1}) = \frac{p(\theta_t \mid \theta_{t-1}, D_{t-1}) \, p(\theta_{t-1} \mid D_{t-1})}{p(\theta_t \mid D_{t-1})} \tag{12.12}$$

Finally, given θ_t and D_{t-1}, the denominator on the right-hand side is the normalising factor, the first term in the numerator on the right-hand side follows from the system equations, and the second term in the numerator is the posterior density at time $t - 1$, which follows from the Kalman filter. If we now assume the distributions are normal we obtain the result

$$\theta_{t-1} \mid D_t \sim N(a_t(-1), R_t(-1)) \tag{12.13}$$

where

$$
\begin{aligned}
a_t(-1) &= m_{t-1} + B_{t-1}(m_t - a_t) \\
R_t(-1) &= C_{t-1} - B_{t-1}(R_t - C_t) B'_{t-1} \\
B_{t-1} &= C_{t-1} G' R_t^{-1}
\end{aligned} \tag{12.14}
$$

You can find more details in Pole et al. (1994).

12.3 Fitting to simulated univariate time series

12.3.1 Random walk plus noise model

As a first example, consider a daily stock price taken at the close of each trading day. This is treated as independent normal random variation, with a standard deviation of 1 about a mean that is 20 for the first 10 time points but drops to 10 for time points 11 up to 20. In practice, we never know the underlying process, and models we fit are based on physical intuition and the goodness-of-fit to the data. State space models have the great advantage that the parameters can change over time and are able to allow for a change in mean level. We first implement a model for the stock price, y_t,

$$y_t = \theta_t + v_t \qquad (12.15)$$
$$\theta_t = \theta_{t-1} + w_t$$

where $\theta_0 \sim N(25,10)$, $v_t \sim N(0,2)$, and $w_t \sim N(0,0.1)$, which allows for small changes in an underlying mean level θ_t. The SS function in the sspir package (Dethlefsen and Lundbye-Christensen, 2006) creates a state space object.[3] The syntax corresponds precisely to the notation of Equation 12.1 except for the additional phi parameter, which we do not use in this chapter and can be ignored. The function kfilter gives the Kalman filter estimate of the state at each time point, given the preceding observations and the observation at that time. The function smooth gives the retrospective estimate of the state at each time given the entire time series.

```
> library(sspir)
> set.seed(1)
> Plummet.dat <- 20 + 2*rnorm(20) + c(rep(0,10), rep(-10,10))
> n <- length(Plummet.dat)
> Plummet.mat <- matrix(Plummet.dat, nrow = n, ncol = 1)
> m1 <- SS(y = Plummet.mat,
           Fmat = function(tt,x,phi) return( matrix(1) ),
           Gmat = function(tt,x,phi) return( matrix(1) ),
           Wmat = function(tt,x,phi) return( matrix(0.1)),
           Vmat = function(tt,x,phi) return( matrix(2) ),
           m0   = matrix(25), C0 = matrix(10)
           )
> plot(m1$y, ylab = "Closing price", main = "Simulated")
> m1.f <- kfilter(m1)
> m1.s <- smoother(m1.f)
> lines(m1.f$m, lty = 2)
> lines(m1.s$m, lty = 3)
```

In Figure 12.1, the Kalman filter rapidly settles around 20 because of the relatively high variance of 10 that we have attributed to our initial inaccurate

---

[3] You will need to download this package from CRAN.

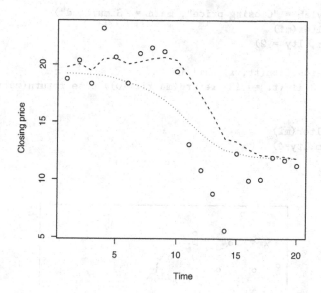

Fig. 12.1. Simulated stock closing prices: filtered estimate of mean level (dashed); smoothed estimate of mean level (dotted).

estimate of 25 until time 10. However, the filter is slower to adapt to the step change. We can improve this performance, if we take advantage of the Bayesian formulation, which is ideally suited for incorporating our latest information. Given the drop in mean level at $t = 11$, it would be prudent to review our assessment of the relevance of the earlier time series to future values. Here, as an example, we decide to assign a variance of 10 to the evolution of θ_t at $t = 12$. This effectively allows the filter to restart at $t = 12$, and in Figure 12.2 you can see that the estimate of θ_t rapidly settles about 10 after $t = 12$. For comparison, the filter without intervention is also shown in Figure 12.2. If you look back at Figure 12.1, you will see the smoothed estimate of θ_t. In this case, it gives less accurate estimates of θ_t than when $t < 20$ because the assumed model, without intervention, makes little allowance for a large step change. In any application, the filter and smoother must coincide for the latest observation. The latest filtered value is our best estimate of the mean level and, for this model, our best estimate of tomorrow's price.

Another means of making the filter adapt more quickly to a change in level is to increase the variance of w_t. The drawback is that the filter will then be unduly influenced by daily fluctuations in price if the level is constant. It is the ratio of the variance of w_t to the variance of v_t rather than the absolute values of the variances that determines the filter path (Exercise 1). However, limits of prediction do depend on the absolute values.

```
plot(m1$y, ylab = "Closing price", main = "Simulated")
m1.f <- kfilter(m1)
lines(m1.f$m, lty = 2)
m2 <- m1
Wmat(m2) <- function(tt, x, phi) {
            if (tt == 12) return(matrix(10)) else return(matrix(0.1))
}

m2.f <- kfilter(m2)
lines(m2.f$m,lty=4)
```

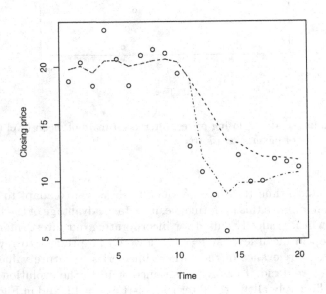

Fig. 12.2. Simulated stock closing prices: filter after intervention (dot-dash); original filter (dashed).

In bull markets, stock prices tend to drift upwards. A drift can be incorporated in the state space model by introducing an additional element in the state vector. You are asked to do this in Exercise 2.

12.3.2 Regression model with time-varying coefficients

The time series regression models that we considered in Chapter 5 are based on an assumption that the process is stationary and hence that the coefficients are constant. This assumption is particularly equivocal for recent environmental and economic time series, even if the predictor variables do not explicitly include time, and state space models are ideally suited to relaxing the restriction. To demonstrate the procedure, we will generate data from the equations

$$y_t = a + bx_t + z_t$$
$$x_t = 2 + t/10$$

(12.16)

where $t = 1, \ldots, 30$, $z_t \sim N(0,1)$, $a = 4$ and $b = 2$ for $t = 1, \ldots, 15$, and $a = 5$ and $b = -1$ for $t = 16, \ldots, 30$. We fit a straight line with time-varying coefficients, which is the model recommended to the sales manager in the house building company of §12.2.1. The components of θ_t are the intercept and slope at time t, and the estimates from the Kalman filter are shown in Figure 12.3. In this application, the matrix F, Fmat in the SS function, is time varying and is $(1, x_t)'$. The matrix Gmat is the identity matrix, which we use diag to generate. The parameter variation, which is modelled with matrix Wmat, is small relative to the observation variance modelled by Vmat. The initial guesses for the intercept and slope are 5 and 3, respectively, and the associated variance of 10 reflects the considerable uncertainty.

```
> library(sspir)
> set.seed(1)
> x1 <- 1:30
> x1 <- x1/10 + 2
> a <- c(rep(4,15), rep(5,15))
> b <- c(rep(2,15), rep(-1,15))
> n <- length(x1)
> y1 <- a + b * x1 + rnorm(n)
> x0 <- rep(1, n)
> xx <- cbind(x0, x1)
> x.mat <- matrix(xx, nrow = n, ncol = 2)
> y.mat <- matrix(y1, nrow = n, ncol = 1)

> m1 <- SS(y = y.mat, x = x.mat,
        Fmat = function(tt,x,phi)
           return( matrix(c(x[tt,1], x[tt,2]), nrow = 2, ncol = 1)),
        Gmat = function(tt,x,phi) return (diag(2)),
        Wmat = function(tt, x, phi) return (0.1*diag(2)),
        Vmat = function(tt,x,phi) return (matrix(1)),
        m0 = matrix(c(5,3),nrow=1,ncol=2),C0=10*diag(2)
        )

> m1.f <- kfilter(m1)
> par(mfcol=c(2,1))
> plot(m1.f$m[,1], type='l')
> plot(m1.f$m[,2], type='l')
```

The estimates from the Kalman filter rapidly approach the known values, even after the step change (Fig. 12.3). The estimated standard errors of the estimates from the filter are based on the specified values in the variance matrices. rather than being estimated from the data (§12.7). Pole et al. (1994) give further details of state space models with estimated variances.

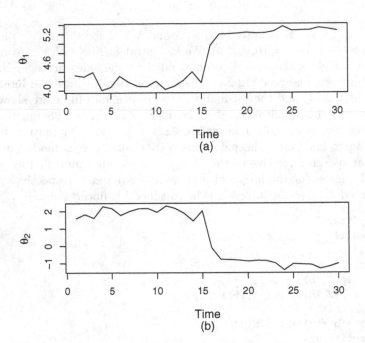

Fig. 12.3. Realisation of a regression model in which the intercept and slope change. Kalman filter estimates of (a) intercept and (b) slope.

12.4 Fitting to univariate time series

Morgan Stanley share prices for trading days from Monday, August 18, 2008, until Friday, November 14, 2008, are in the online file `MorgStan.dat`. If we wish to set up a random walk plus drift model, we need an estimate of the two variance components. One way of doing this is to compare the variances within and between weeks, the former being taken as V and the latter as W (Fig. 12.4). Late 2008 was a tumultuous period for bank shares and the variances within and between weeks are estimated as 5.4 and 106, respectively (Exercise 3). With these parameter values, both the filtered and smoothed values are very close to the observed data. The estimated price of shares on Monday, November 17, is given by the latest filtered value `m1.f$m[n]` and equals 12.08, which is close to the 12.03 closing price on Friday, November

14. If the variances within and between weeks are estimated from the first four weeks, when the market was relatively stable, they are 2.1 and 1.0. The estimated price of shares on Monday, November 17, is now 12.69.

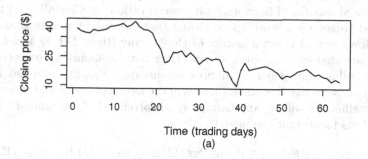

Time (trading days)
(a)

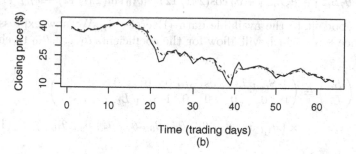

Time (trading days)
(b)

Fig. 12.4. Morgan Stanley close of business share prices for trading days August 18 until November 14, 2008. Kalman filter and smoothed values: (a) $V = 5.4$ and $W = 106$; (b) $V = 2.2$ and $W = 1.0$.

12.5 Bivariate time series – river salinity

The Murray River supplies about half of South Australia's urban water needs and, in dry years, this can increase to as much as ninety percent. Other sources of water in South Australia are bore water and recycled water, although both tend to have high salinity. The World Health Organisation (WHO) recommendation for the upper limit of salinity of potable water is 500 mg/l (approximately 800 EC), but the domestic grey water system and some industrial and irrigation users can tolerate higher levels of salinity. The low rainfall and increasing population makes the efficient use of water resources a priority, and there are water-blending schemes that aim to maximise the use of recycled water. The average monthly salinity, measured by electrical conductivity (EC; microSiemens per centimetre at 25°C) and flow (Gigalitres per month) at

Chowilla on the Murray River, have been calculated from data provided by the Government of South Australia and are available in the file `Murray.txt`. Predictions of salinity are needed for the recycled water schemes to be operated efficiently, and the changing level of salinity requires an adaptive forecasting strategy, which is well supported by state space methods.

We have 81 months of flows and salt concentrations at Chowilla and have been asked to set up a state space model that can adapt to the changing salt and flow levels that are a feature of the Murray River. Let S_t and L_t be the mean adjusted salt concentration and river flow for month t, respectively. The mean adjustment generally improves numerical stability and has as a convenient consequence that the estimates of the intercept terms and estimates of other coefficients will be approximately uncorrelated. A preliminary time series analysis found that the model

$$S_t = \theta_1 + \theta_2 S_{t-1} + \theta_3 L_{t-1} + \theta_4 \cos(2\pi t/12) + \theta_5 \sin(2\pi t/12) + v_{S,t} \quad (12.17)$$

$$L_t = \theta_6 + \theta_7 S_{t-1} + \theta_8 L_{t-1} + \theta_9 \cos(2\pi t/12) + \theta_{10} \sin(2\pi t/12) + v_{L,t} \quad (12.18)$$

provides a good fit to the available data (Exercise 4). We now express this in state space form, which will allow for the coefficients $\theta_1, \ldots, \theta_{10}$ to change over time.

$$\begin{pmatrix} S_t \\ L_t \end{pmatrix} = \begin{pmatrix} 1 & S_{t-1} & L_{t-1} & cs_t & sn_t & 0 & 0 & 0 & 0 & 0 \\ 0 & 0 & 0 & 0 & 0 & 1 & S_{t-1} & L_{t-1} & cs_t & sn_t \end{pmatrix}$$

$$\times \begin{pmatrix} \theta_{1,t} & \theta_{2,t} & \theta_{3,t} & \theta_{4,t} & \theta_{5,t} & \theta_{6,t} & \theta_{7,t} & \theta_{8,t} & \theta_{9,t} & \theta_{10,t} \end{pmatrix}' \quad (12.19)$$

$$\theta_{i,t} = \theta_{i,t-1} + w_{i,t} \quad (12.20)$$

The R code is now more succinct and makes use of `diag` to set up diagonal matrices, but the general principles are the same as for any regression model with time-varying coefficients. The diagonal elements of the matrix `Vmat` are the estimated variances of the errors from the preliminary regression models for S_t and L_t, which are 839 and 1612, respectively. There is no evidence of autocorrelation in the residual series from the two regressions, but the cross-correlation at lag 0, which is -0.299, is statistically significant. The corresponding estimate of the covariance of the errors, the off-diagonal term in `Vmat`, is therefore -348. The matrix `Wmat` is set up to allow the mean levels to adapt and slight adaptation of the other coefficients. The choice of values for the variances is somewhat subjective. The mean salinity and mean flow over the 81-month period were 165 and 259, respectively, and a variance of 10 corresponds to a standard deviation of roughly 2% of mean salinity and 1% of mean flow. The variances of 0.0001 correspond to a standard deviation of 0.01 for the change in level of the other coefficients. The effect of changing entries in `Wmat` can be investigated when setting up the filter. The initial estimates in `m0` were set fairly close to the estimates from the preliminary regressions. The

uncertainty associated with these estimates was arbitrarily set at 100 times
Wmat, as it does not have a critical effect on the performance of the filter. The
results are shown in Figures 12.5 and 12.6 and seem reasonable.

```
> library(sspir)
> www <- 'http://www.massey.ac.nz/~pscowper/ts/Murray.txt'
> Salt.dat <- read.table(www, header=T) ; attach(Salt.dat)
> n <- 81 ; Time <- 1:n
> SIN  <- sin(2 * pi * Time /12)[-1]
> COS  <- cos(2 * pi * Time /12)[-1]
> Chowilla <- Chowilla - mean(Chowilla)
> Flow <- Flow - mean(Flow)
> Chow <- Chowilla[-1]
> Chow.L1 <- Chowilla[-n]
> Flo <- Flow[-1]
> Flo.L1 <- Flow[-n]
> Sal.mat <- matrix(c(Chow, Flo), nrow = 80, ncol = 2)
> x0 <- rep(1, (n-1))
> xx <- cbind(x0, Chow.L1, Flo.L1, COS, SIN)
> x.mat <- matrix(xx, nrow = n-1, ncol = 5)
> G.mat <- diag(10)
> W.mat <- diag(rep(c(10, 0.0001, 0.0001, 0.0001, 0.0001), 2))
> m1 <- SS(y = Sal.mat, x = x.mat,
          Fmat =
            function(tt, x, phi) return (matrix(
              c(x[tt,1], x[tt,2], x[tt,3], x[tt,4], x[tt,5], rep(0,10),
                x[tt,1], x[tt,2], x[tt,3], x[tt,4], x[tt,5]),
                                nrow=10,ncol=2)),
          Gmat = function(tt, x, phi) return (G.mat),
          Wmat = function(tt, x, phi) return (W.mat),
          Vmat = function(tt, x, phi) return
                    (matrix(c(839, -348, -348, 1612), nrow=2, ncol=2)),
          m0=matrix(c(0,0.9,0.1,-15,-10,0,0,0.7,30,20),nrow=1,ncol=10),
          C0 = 100 * W.mat
          )

> m1.f <- kfilter (m1)
> par(mfcol=c(2,3))
> plot(m1.f$m[,1], type='l')
> plot(m1.f$m[,2], type='l')
> plot(m1.f$m[,3], type='l')
> plot(m1.f$m[,6], type='l')
> plot(m1.f$m[,7], type='l')
> plot(m1.f$m[,8], type='l')
>
> par(mfcol=c(2,2))
> plot(m1.f$m[,4], type='l')
> plot(m1.f$m[,5], type='l')
```

```
> plot(m1.f$m[,9], type='l')
> plot(m1.f$m[,10], type='l')
```

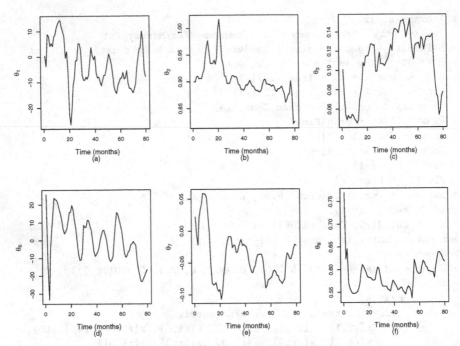

Fig. 12.5. Kalman filter estimates of parameters for the Murray River salt and flow model: (a) deviation of salt from mean; (b) autoregressive coefficient for salt; (c) cross regressive coefficient for flow; (d) deviation of flow from mean; (e) cross regressive coefficient for salt; (f) autoregressive coefficient for flow.

12.6 Estimating the variance matrices

In the applications considered in this chapter, we have had to specify values for the V and W matrices when setting up the state space model. We can adapt the algorithm to update these values as we obtain more data. The formula for the variance of one-step-ahead forecast errors depends on the known values of entries in the V and W matrices. This theoretical variance can be compared with the variance of the actual forecast errors up until time t. Define

$$\phi_t = \frac{\text{actual variance of forecast errors up to time } t}{\text{theoretical variance of forecast errors}}$$

Then updated matrices for V and W are obtained from $\phi_t V$ and $\phi_t W$, respectively. This strategy will not adjust the relative variances of measurement noise and system noise, but it does allow the absolute values to be updated (Exercise 7).

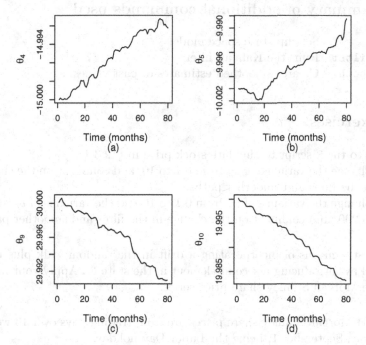

Fig. 12.6. Kalman filter estimates of coefficients of seasonal components for Murray River salt and flow model: (a) cosine for salt; (b) sine for salt; (c) cosine for flow; (d) sine for flow.

12.7 Discussion

One of the main advantages of state space models is that they are adaptive, and the benefits of this are usually realised by implementing them in real time. We have only covered relatively straightforward examples, and there are many useful extensions. In particular, there are sophisticated methods for estimating the variances rather than specifying them and methods for estimating parameters in the F and G matrices as well as the states when this is theoretically possible. The distinction between states and unknown parameters depends on the application (see Exercise 5).

The Kalman filter applies to linear systems, but it can also be used as a local linear approximation to a non-linear system. This important development is known as the extended Kalman filter. The optimality of the standard Kalman filter rests on an assumption that the noise is Gaussian (normal) and independent, but similar optimal filters have been developed with other assumptions about noise distributions.

12.8 Summary of additional commands used

SS	Sets up state space model
kfilter	Runs the Kalman filter
smooth	Creates smoothed estimates of past values

12.9 Exercises

1. Refer to the R script to simulate stock price in §12.3.1.
   a) Change the variance of w_t from 0.1 to 10, and comment on the change in the filter and smoother paths.
   b) Change the variance of w_t from 0.1 to 10 and the variance of v_t from 2 to 200, and comment on the change in the filter and smoother paths.

2. Suggest a means of incorporating a drift in the random walk plus noise model by introducing a second element in the state θ_t. Apply your model to the Morgan Stanley share price series.

3. The 64 Morgan Stanley share prices are from trading days over 13 weeks, Monday, September 1, being the Labor Day holiday.
   a) Calculate the variance for each of the 13 weeks, all but the third consisting of five trading days. This is the estimate of the within-week variance S^2_{within}.
   b) Calculate the mean for each of the weeks and calculate the variance of the 13 means. Denote this by $S^2_{\bar{x}}$.
   c) Estimate the variance between weeks as

   $$S^2_{between} = S^2_{\bar{x}} - \frac{1}{5}S^2_{within}$$

   The slight inaccuracy that results from one week having only four trading days is negligible.
   d) If you are familiar with the analysis of variance, use aov to obtain estimates of the within-week variances that do allow for the Labor Day holiday.

4. Calculate the preliminary regressions for the Murray River salinity example, and verify the numbers given in the text.

5. In many control applications, the matrices F_t and G_t are constant. The first issue is whether or not all the values of the state at time t can be inferred from the observations at time t. If they can be inferred, the system is said to be observable. The linear system in Equation (12.1) is observable if the observability matrix, O, defined by

$$O = \left(F' \ F'G \cdots F'G^{p-1} \right)'$$

has full rank, p. Consider a state space model for mean adjusted salinity (S_t) and flow (L_t) in which the coefficients are assumed known and the salinity and flow are the components of the state. Suppose only flow is measured, and assume the model has the form

$$L_t = (0 \ \ 1) \left(S_t \ \ L_t \right)' + v_t$$

$$\begin{pmatrix} S_t \\ L_t \end{pmatrix} = \begin{pmatrix} a & b \\ c & d \end{pmatrix} \begin{pmatrix} S_{t-1} \\ L_{t-1} \end{pmatrix} + \begin{pmatrix} w_{S,t} \\ w_{L,t} \end{pmatrix}$$

Can the salinity at time t be inferred from the flow measurement? If so, what are the conditions on a, b, c, and d? (In time series applications, F_t will generally contain the values of the predictor variables at time t. The observability requirement at time t is that the $t \times 1$ columns of predictors in the linear regression model be linearly independent.)

6. Suppose that an observation y has a normal distribution with mean θ and variance ϕ and that the prior distribution for θ is normal with mean θ_0 and variance ϕ_0. We require the posterior distribution of θ given the observation y. From Bayes's Theorem,

$$p(\theta \,|\, y) \propto p(\theta) \, p(y \,|\, \theta)$$

$$\propto \exp \left[-\frac{(\theta - \theta_0)^2}{2\phi_0} \right] \exp \left[-\frac{(y - \theta)^2}{2\phi} \right]$$

It is now convenient to anticipate the result, that the mean of the posterior distribution (θ_1) is a weighted mean of the mean of the prior distribution and the observation with weights proportional to the precision, and define θ_1 and ϕ_1 as

$$\theta_1 = \frac{\phi_0^{-1}}{\phi_0^{-1} + \phi^{-1}} \theta_0 + \frac{\phi^{-1}}{\phi_0^{-1} + \phi^{-1}} x$$

$$\phi_1^{-1} = \phi_0^{-1} + \phi^{-1}$$

Now use these expressions to replace θ_0 and ϕ_0 in the expression for $p(\theta \,|\, y)$ by θ_1 and ϕ_1:

$$p(\theta \,|\, y) \propto \exp \left[-\frac{\theta^2 + 2\theta\theta_1}{2\phi_1} \right]$$

This is proportional to a normal distribution with mean θ_1 and variance ϕ_1 since θ_1^2/ϕ_1 is a constant with respect to the prior distribution. So,

$$p(\theta, y) = \frac{1}{\sqrt{2\pi\phi_1}} \exp \left[-\frac{(\theta - \theta_1)^2}{2\phi_1} \right]$$

The extension of this result to the multivariate normal distribution can be used to derive the Kalman filter.

7. Verify that the following script is a Kalman filter. Compare its performance with **kfilter** on the simulated regression example. Adapt the filter to update the variance and investigate its performance.

```
> set.seed(1)
> x1 <- c(1:30)
> x1 <- x1/10 + 2
> a <- 4
> b <- 2
> n <- length(x1)
> y1 <- a + b * x1 + 0.1 * rnorm(n)
> x0 <- rep(1, n)
> xx <- cbind(x0, x1)
> F <- matrix(xx, nrow = n,ncol=2)
> y <- matrix(y1, nrow = n,ncol=1)
> G <- matrix(c(1,0,0,1), nrow = 2, ncol = 2)
> W <- matrix(c(1,0,0,1), nrow = 2, ncol = 2)
> V <- matrix(1)
> m0 <- matrix(c(5,1.5), nrow = 2, ncol = 1)
> C0 <- matrix(c(.1,0,0,.1), nrow = 2, ncol = 2)
> a <- 0;R <- 0;f <- 0;Q <- 0;e <- 0;A <- 0;m <- 0;C <- 0;tt <- 0;
> Kfilt.m <- cbind(rep(0, n), rep(0, n))
> m <- m0
> C <- C0
> for (tt in 1:n) {
    Fmat <- matrix(c(F[tt,1],F[tt,2]), nrow = 2, ncol = 1)
    a <- G %*% m
    R <- G %*% C %*% t(G) + W
    f <- t(Fmat) %*% a
    Q <- t(Fmat) %*% R %*% Fmat + V
    e <- y[tt]-f
    A <- R %*% Fmat %*% solve(Q)
    m <- a + A %*% e
    C <- R - A %*% Q %*% t(A)
    Kfilt.m[tt,1] <- m[1,1]
    Kfilt.m[tt,2] <- m[2,1]
  }
> plot(Kfilt.m[1:n, 1])
> plot(Kfilt.m[1:n, 2])
```

References

Akaike, H. (1974). A new look at statistical model indentification. *IEEE Transactions on Automatic Control*, 19:716–722.

Bass, F. (1969). A new product growth model for consumer variables. *Management Science*, 15:215–227.

Becker, R., Chambers, J., and Wilks, A. (1988). *The NEW S Language*. New York: Chapman & Hall.

Beran, J. (1994). *Statistics for Long-Memory Processes*. New York: Chapman & Hall.

Bozdogan, H. (1987). Model selection and Akaike's Information Criterion (AIC): The general theory and its analytical extensions. *Psychometrika*, 52:345–370.

Brohan, P., Kennedy, J., Harris, I., Tett, S., and Jones, P. (2006). Uncertainty estimates in regional and global observed temperature changes: A new data set from 1850. *Journal of Geophysical Research*, 111(D12106).

Brown, R. (1963). *Smoothing, Forecasting, and Prediction of Discrete Time Series*. Englewood Cliffs, NJ: Prentice-Hall.

Coleman, S., Gordon, A., and Chambers, P. (2001). SPC - making it work for the gas transportation business. *Journal of Applied Statistics*, 28:343–351.

Colucci, J. and Begeman, C. (1971). Carcinogenic air pollutants in relation to automotive traffic in New York. *Environmental Science and Technology*, 5:145–150.

Dalgaard, P. (2002). *Introductory Statistics with R*. New York: Springer.

Dethlefsen, C. and Lundbye-Christensen, S. (2006). Formulating state space models in R with focus on longitudinal regression models. *Journal of Statistical Software*, 16(1):1–15.

Enders, W. (1995). *Applied Economic Time Series*. New York: John Wiley & Sons.

Holt, C. (1957). Forecasting trends and seasonals by exponentially weighted averages. *O.W.R. Memorandum no.52, Carregie Institute of Technology*.

Hurst, H. (1951). Long-term storage capacity of reservoirs. *Transactions of the American Society of Civil Engineers*, 116:770–779.

P.S.P. Cowpertwait and A.V. Metcalfe, *Introductory Time Series with R*, 247
Use R, DOI 10.1007/978-0-387-88698-5_BM1,
© Springer Science+Business Media, LLC 2009

Iacobucci, A. and Noullez, A. (2005). A frequency selective filter for short-length time series. *Computational Economics*, 25:75–102.

Ihaka, R. and Gentleman, R. (1996). R: A language for data analysis and graphics. *Journal of Graphical and Computational Statistics*, 5:299–314.

Jones, P. and Moberg, A. (2003). Hemispheric and large-scale surface air temperature variations: An extensive revision and an update to 2001. *Journal of Climate*, 16:206–223.

Kalman, R. and Bucy, R. (1961). New results in linear filtering and prediction theory. *Journal of Basic Engineering*, 83:95–108.

Leland, W., Taqqu, M., Willinger, W., and Wilson, D. (1993). *On the self-similar nature of ethernet traffic.*

Mahajan, V., Muller, E., and Wind, Y. (eds.) (2000). *New-Product Diffusion Models.* New York: Springer.

Metcalfe, A., Maurits, L., Svenson, T., Thach, R., and Hearn, G. (2007). Modal analysis of a small ship sea keeping trial. *ANZIAM Journal*, 47:915–933.

Mudelsee, M. (2007). Long memory of rivers from spatial aggregation. *Water Resources Research*, 43(W01202).

Nason, G. (2008). *Wavelet Methods in Statistics with R.* New York: Springer.

Perron, P. (1988). Trends and random walks in macroeconomic time series. *Journal of Economic Dynamics and Control*, 12:297–332.

Pfaff, B. (2008). *vars: VAR Modelling.* R package version 1.3-7.

Plackett, R. (1950). Some theorems in least squares. *Biometrika*, 37:149–157.

Pole, A., West, M., and Harrison, J. (1994). *Applied Bayesian Forecasting.* New York: Chapman & Hall.

Rayner, N., Parker, D., Horton, E., Folland, C., Alexander, L., Rowell, D., Kent, E., and Kaplan, A. (2003). Globally complete analyses of sea surface temperature, sea ice and night marine air temperature, 1871-2000. *Journal of Geophysical Research*, 108(4407).

Rogers, E. (1962). *Diffusion of Innovations.* New York: Free Press.

Romilly, P. (2005). Time series modelling of global mean temperature for managerial decision making. *Journal of Environmental Management*, 76:61–70.

Said, S. and Dickey, D. (1984). Testing for unit roots in autoregressive moving average models of unknown order. *Biometrika*, 71:599–607.

Siau, J., Graff, A., Soong, W., and Ertugrul, N. (2004). Broken bar detection in induction motors using current and flux spectral analysis. *Australian Journal of Electrical & Electronics Engineering*, 1:171–177.

Thomson, W. (1993). *Theory of vibration with applications*, 4th Edition. London: Chapman & Hall.

Trapletti, A. and Hornik, K. (2008). *tseries: Time Series.* R package version 0.10-14.

Winters, P. (1960). Forecasting sales by exponentially weighted moving averages. *Management Science*, 6:324–342.

Index

acceleration series, 186
acf, 35, 47, 73
adf.test, 214
aggregate, 7, 17
aggregation, 7
AIC, 41, **84**, 144
air passengers, 4, 62
aliasing, 180
ar, 82, 84, 124, 218, 221
AR model, 79
    Exchange Rates, 84
    Global Temperatures, 85
    stationary, 79
ARCH
    definition, 147
ARCH model, 145
arima, 124, 138, 140
ARIMA model, 137
    definition, 139
    fractional, 160
    seasonal, 142
arima.sim, 129, 140
ARMA model, 127
as.vector, 13
Australian wine sales, 60
autocorrelation, 33
    function, 33
    partial, 81
autocovariance
    function, 33
autoregressive model, 79

autoregressive moving average, 127

backward shift operator, 71, 79
bandwidth, 177
bank loan rate, 166, 189
Bass model, 51, 52
Bayes's Theorem, 231
beer production, 10, 141
Bellcore Ethernet, 165
bias correction, 115
bivariate time series, 239
bivariate white noise, 219
bootstrapping, 68, 222
boxplot, 7
building activity, 46, 47
business cycles, 6

cbind, 11
ccf, 48, 49, 220
characteristic equation, 79, 142, 221
chocolate production, 10
class, 4
climate change, 1, 16
coef, 94
cointegration, 216
conditional heteroskedastic, 137, 147
confidence intervals, 99
confint, 96, 99
cor, 30
correction factor, 117
correlation, 2, 13, 30, 33
    autocorrelation, 33
    causation, 13

P.S.P. Cowpertwait and A.V. Metcalfe, *Introductory Time Series with R*,   249
Use R, DOI 10.1007/978-0-387-88698-5_BM2,
© Springer Science+Business Media, LLC 2009

Printed in the United States
By Bookmasters